BEES IN AMERICA

BEES IN AMERICA

How the Honey Bee Shaped a Nation

TAMMY HORN

THE UNIVERSITY PRESS OF KENTUCKY

Publication of this volume was made possible in part by a grant
from the National Endowment for the Humanities.

Scholarly publisher for the Commonwealth,
serving Bellarmine University, Berea College, Centre
College of Kentucky, Eastern Kentucky University,
The Filson Historical Society, Georgetown College,
Kentucky Historical Society, Kentucky State University,
Morehead State University, Murray State University,
Northern Kentucky University, Transylvania University,
University of Kentucky, University of Louisville,
and Western Kentucky University.

Editorial and Sales Offices: The University Press of Kentucky
663 South Limestone Street, Lexington, Kentucky 40508-4008
www.kentuckypress.com

05 06 07 08 09 5 4 3 2 1

Library of Congress Cataloging-in-Publication Data
Horn, Tammy, 1968-
Bees in America : how the honey bee shaped a nation / Tammy Horn.
p. cm.
Includes bibliographical references and index.
ISBN 0-8131-2350-X (hardcover : alk. paper)
1. Bee culture—United States—History. 2. Honeybee—
United States—History. I. Title.

SF524.5.H67 2005
638'.1'0973—dc22
2004026887

This book is printed on acid-free recycled paper meeting
the requirements of the American National Standard
for Permanence in Paper for Printed Library Materials.

Manufactured in the United States of America.

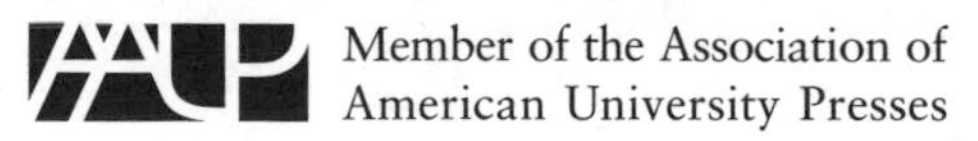

To a strong colony:

Earl and Charlene Horn,
Anna Lee Hacker,
Jennifer Peckinpaugh,
Scottie Noland,
Carol Falkenstine de Rosset

Contents

ILLUSTRATIONS

Acknowledgments

William Butler Yeats once said one must labor to be beautiful. I no longer deny I am a worker bee. I like the activity and anonymity that happens when I am immersed in a project I enjoy. Of all the bees, my favorite is the forager, the bee that finds flowers in the fields, collects dusty pollen and nectar, and carries these raw materials back to the hive to be turned into nourishment for an entire colony.

I have been surrounded by a symphony of support, many of whom labor anonymously to make our society a better place. In addition to those mentioned on the dedication page, I want to acknowledge the various social families that have dreamed, written, and contributed to this book.

The lines between relatives and friends have blurred so long ago that it is impossible for me to name all the people I consider family. Together with my parents, my immediate family—Jamie Horn, Brian and Gabriel Napier, Lyn, and E. J. Hacker—have been unanimous in their support of my goals. Amy Noland Hughes and her family—Joel Scott, Joy Lee, Cathy, Julie, Keith, Becky, Jay, Tracy, Keith, Becky, Sue, Nancy, and Eddie—have provided steady, consistent friendship.

Forever etched in my mind is the kindness of Ashley Gibson Khazen and her family, Sally, Merry, Tom, Gloria, and Haidar. I have needed the sage advice of Debra Lewis, Bruce Danner, Chris and Kateri Chambers—their words filtered through jazz, bourbon, and Pontchartrain humidity. Jennifer Lewis, Francis Figart, Jim Kenkel, and Sherry Robinson have continued to serve as loci of value. And great gratitude is extended to Emily Saderholm and Andy Teague, whose impromptu kindness and steady strength I have taken great comfort in. Their introduction to

Hawaii—its emphasis on peace and acceptance—sustained me during the composition process. All of these people—such strong convictions, such gentle hearts—have helped me find my own words.

My colleagues have provided solid sources of support. At Fort Hays, Sharon Wilson, Kris Bair, Jay Osiovitch, Cliff Edwards, and Ralph Voss were wonderful teachers. I extend thanks to those at the University of Alabama—Ralph Voss (again), Phil Beidler, Rich Megraw, Richard Rand, Diane Roberts, John and Amy Beeler, Larry and Maureen Kohl, Kathy and Paul Gorman, Dwight Eddins, Jennifer Horn, Dave Johnston, Pat Hermann, Alan Wier, Tony and Jessica Brusate, Neil and Janet Kirchner, Jim Salem, Rose Gladney, and Lyn Adrian.

I don't want to forget my colleagues and carpool group at the University of West Alabama, who encouraged me to follow through with my larger dreams. Mitzi Forrester Gates, Tina Naremore Jones, Mark Griffith, Richard Schelhammer, Robyn Trippany, Henry Walker, Leesa Corrigan, and Roy Underwood were attentive listeners and storytellers. Many a mile melted away in laughter and wisdom among the carpool. I remain grateful for their forthrightness.

My colleagues at Eastern Kentucky University warmly welcomed me back to Kentucky: Jen Spock, Deb Core, Joan Miller, Belinda Gadd, Charlie Sweet, Hal Blyth, Debbie Whalen, Carrie Cooper, Bonnie Plummer, Dan Florell, Kevin Jones, Martha Marcum, Kevin Rahimzadeh, Sarah Tsiang, Anne Gossage, Susan Kreog, Meg Matheny, Jennings Mace, and all the busy bees on the second floor.

No doubt about it—this book is a dream come true, in large part because it was written at my alma mater, Berea College. I would like to thank Eugene Startzman, Libby Jones, Deanna Sergel, Jackie Burnside, Richard Sears, Al DeGiacomo, Jim Gage, Beth Crachiolo, Deb Martin, Steve Pulsford, Stephanie Browner, Dave Porter, Laura Crawford, Barbara Wade, Randall Roberts, Fred de Rosset, John Carlevale, Sandy Bolster, Beth Curlin, Patty Tarter, Francie Bauer, Barbara Powers, Ann Chase, Jill Bouma, Karl Walhausser, Bill Ramsey, Cary Hazelwood, Kathryn Akural, Susan Vaughn, Linda Varwig, Don Hudson, Gary Mahoney, Duane Smith, Dave Bowman, Rob Foster, Rebecca Bates, Carolyn Castle, Cindy Judd and Susan Henthorne, and especially Phyllis Gabbard, to name just a few. The Berea community has extended the campus in a nice, quiet way, continually offering encouragement, especially in the smiles of Praveena Salins, Gary Elam, John-David Startzman, Charlotte Hazeltine, and Monica Isaacs. My students at Berea College and Eastern Kentucky University, and my teaching assistant Jaime Breckenridge, have been good-natured recipients of my research. Although Thomas Wolfe expressed doubt that one could go home again, I have experienced profound joy in doing so.

When one works with bees, one quickly realizes the importance of dances. Special thanks go to the Lexington and Berea dance communities: Cary Ravitz; Fran Bevins; Richard King; Liz Donaldson; Jackie, John, and Eric Crowden; Barbara Ramlow; Barbara Lytle; Yoong-Geum Ahm; Teresa Cole; Steve Bennett; Lucinda Masterson; Lindsey and Olivia Morris; Bob Lovett; Hannah Kirsch; Kevin Hopper; Trent Ripley; Dan Van Treese; Joe Carwile; and Larry Johnson.

Gratitude goes to my church family: Jean and Bob Boyce, Mary Lou and Les Pross, Bob and Liz Menefee, Kent Gilbert and Jan Pearce, Marlene and John Payne, Dorothy and Gene Chao, Harry Rice and Carol Gilliam, Eddie Broaddus and Loretta Reynolds, Sean Perry, Robert Rorrer, Jan Hamilton, Al and Alice White, Matt Saderholm and Angela Anderson, John and Ramona Culp, John and Keila Thomas, Susan Yorde, and Nancy and Larry Shinn.

The Bee Project: The University Press of Kentucky initiated this project, and their open-door policy has been much appreciated by this writer. Steve Wrinn, Gena Henry, Joyce Harrison, and Craig Wilkie have been patient and enthusiastic. I never doubted that I was in good hands with the marketing department: Leila Salisbury, Wyn Morris, Allison Webster, and Mack McCormick. But I am especially appreciative to those who worked with me in the editorial process, which is so much more than just cleaning up syntax. Freelancer Karen Hellekson provided tactful editorial suggestions and revisions. Danielle Dove carefully designed the text in her corner of the Press. Most of all, I have appreciated Nichole Lainhart, who has coaxed this manuscript and me along with smiles and an invaluable sense of humor.

Thanks goes to Kyle McQueen, whose conscientious research of nineteenth-century newspaper articles helped spice up those chapters, and Lowell Bouma, who translated Carl Seyffert's 1930 German text into English. Utah deserves special mention for its generous archivists. Carol Edison, Bonnie Lee Sparks, and Sharon Odekirk provided information, materials, and warm personalities. Of course, the archivists throughout the states responded with enthusiasm and good will: Cathy Grosfils, Richard Doty, Dot Wiggins, Cathy Michelini, Brian Thompson, Larrie Curry, Christian Goodwillie, Doug Nesbitt, Frank Smith, Jason Wilson, the National Honey Board, and Almond Board of California were especially generous in providing permissions, suggestions, reminders, etc.

The beekeepers themselves rallied to my project without hesitation. There is a Buddhist saying: when the student is ready, the teacher will be there. Fortunately for me, the teacher was Tom Webster, Kentucky state entomologist. Tom introduced me to editors Joe Graham, Jerry Hayes, and Kim Flottum, all of whom opened their hearts, their archives, and even the bank vault to help me in my research. As if that weren't enough, Tom and Kim patiently read this book in all of its various stages.

A special thanks should be extended to Etta Thacker, Bill Mares, Wyatt Mangum, Eric Mussen, John Harbo, Marla Spivak, Robin Mountain, Phil Craft, Bob and Yvonne Koehnen, John and Jay Miller, Kim Lehman, Suzanne Doerfield, and Sue Cobey, who generously provided impromptu bee lessons in the areas where instruction was needed. If there are any flaws in the book, the fault is mine. Gene Kritsky kindly permitted me to use his illustrations, thus providing clarity to some of the descriptions about hives. And finally, Sarah Minion and Mary Kay Franklin at the Walter Kelley Bee Company provided resources, stories, pictures, and best of all, memories of both Kelley and my grandfather.

The final two beekeepers, Ted Hacker and Bess Horn, are in a class of their own for more than one reason.

Part One

Hiving Off from Europe

Introduction

But are there not more than enough bee books?

—*Karl von Frisch*

Von Frisch's question has haunted me throughout the process of compiling this book. For those interested in how to keep bees, many fine writers already exist. For those who want to read about the joys of beekeeping, better books than this one are already on the market. Even scientists and researchers have found an appreciative general audience. Von Frisch decided to "give the reader the interesting part of the subject, without the ballast of practical instruction." The result is *The Dancing Bees;* he won the Nobel Prize for his life's work in honey bee communication in 1973.

Now, I offer another bee book, better defined by what it is *not* than what it *is*. Absent are the latest statistics about honey, beeswax, or imports. Nor will this book prepare anyone to don a veil, grab a smoker, and head for the nearest bee tree. I am not a biologist.

So why keep reading? The answer is my desire to examine the values associated with being an American, as complicated as that definition can be. No two values have been so highly regarded since colonial days than industry and thrift. No better symbol represents these values than the honey bee. Furthermore, although almost no part of our culture remains untouched by honey bees, the field of cultural entomology is still relatively unexplored. Writing as recently as 1987, Charles Hogue lamented of honey bees that "[the bees] cultural importance relative to that of other life forms is not known, because a comparative study has not yet been conducted."[1] Eva Crane has since responded with *The World History of Beekeeping and Honey Hunting* (1999), which I have relied upon through the course of writing this book. Much is left out, however, for it seems to me that a crucial discussion about the interactions between the beekeepers themselves needs to be added to the literature. I look at how four elements at an intersection called America affect beekeeping in irrevocable ways; those four elements are the honey bee itself, the ideas that Americans had about honey bees, the freedom to develop those ideas, and the beekeepers' interactions with each other. Of these, it is the last I find so very compelling, for I have a hunch that when beekeepers form a society, they in turn affect the larger society we call America in fascinating ways.

Why do we associate industry and thrift with the honey bee? One glance in a hive clarifies why we consider bees to be the most industrious of insects: the bee society is the most perfectly engineered social sphere. Honey bees do not waste an inch of space, honey, or wax in their hives. Their cells are built on a slight upward incline, thus using gravity so that all the honey stays within the cell. Honey bees do not waste time: they have clearly defined tasks that have evolved over millions of years to create a highly structured social system. Furthermore, no matter their location, bees build these hives to protect themselves through the seasons.

America is the exact opposite. At first glance, we are anything but organized. We do not have an official religion, political party, language, or even family structure. The Declaration of Independence assures this country that independence will be a characteristic of American culture. But Americans love successes, especially financial ones. And for many people, the traits associated with honey bee society—industry and thrift—were directly associated with the benefits that the New World offered those European immigrants willing to work hard, take advantage of its natural resources, and save their money until they could buy (or take) land.

Although the honey bee did not officially arrive in America until the 1620s, its image had been associated with America much earlier. As soon as Columbus became convinced that he had arrived near the original

Garden of Eden in 1492, North America quickly became known as the New World, and by extension in the European mind-set, a new Canaan, a land of milk and honey. Even though honey bees and cattle were not native to North America, as soon as colonists imagined that America could be a "land of milk and honey," they set in motion the events to make America so. Behind this determination was a set of complex reasons—political, cultural, and theological—that began in England, but by the end of the century involved Holland, Sweden, France, Germany, and Africa.

Although the book focuses on America, this introduction begins in seventeenth-century England, for in the years preceding Jamestown's foundation, the English established an important social metaphor associated with honey bees that deviated from classical bee metaphors. Having flourished under Queen Elizabeth I, England had continued to invest the honey bee image with values of stability, responsibility, and industry inherited from Roman and Greek writers. But immediately after Queen Elizabeth I died, England experienced three major upheavals: weather disasters, overpopulation, and land transfers as society abandoned feudal policies. The complexities of these changes resulted in high rates of poverty, but the royal authorities under King James and King Charles I quickly adopted a relatively new biological metaphor to simplify the complex social problems they had. The queen's beekeeper, Charles Butler, defined the male honey bee as a drone in 1609; this new distinction in honey bee society offered a simple biological reason for a complex social one. The poor people were labeled "drones."

When English clergy and politicians adopted a beehive metaphor to explain seventeenth-century social ills, they unwittingly initiated complex implications for the new century, the New World, and its new inhabitants, especially those from Africa. The book examines the various communities that arrived in America and their reasons for doing so: religious intolerance, political instability, or land shortages. In addition to Massachusetts, Maryland, and Virginia, other European colonists in the New World, such as those in New Netherlands, New Sweden, and France, brought beekeeping skills with them. These European colonists brought an agrarian philosophy that depended on cows and bees to extend and define an agricultural legacy inherited from classical Greek and Roman writers. The pastoral legacy of bees and cows, it was assumed, would guarantee the New World's potential to provide for the immigrants in search of new opportunities.

But the New World also needed new labor. As much as the colonists valued their own industry, they were not opposed to using slaves. Because the Dutch were extensively involved with the African slave trade, Africa too played an important role in how America implemented its value system. The African tribes valued bees and cows, and this book would be incomplete without some acknowledgment that African American slaves came from prolific beekeeping countries. In short, a basic argument of this book is that America's beekeepers and honey hunters formed a global network as early as the seventeenth century.

Before sugar had become an established product in the Caribbean or North America, bees fulfilled an important need in English diet, economy, and culture. Beeswax and honey were staples of medieval and Renaissance life. Bees provided sweeteners, wax for candles and waterproofing, and honey for mead. In fact, mead was Queen Elizabeth I's favorite drink. And at the opposite end of the social spectrum, peasants used products from the hive to pay taxes, to supplement their diets, and to barter for wheat and salt.[2]

Although the English always had been passionate about their bees, seventeenth-century writers were really creative in their admiration of honey bees. Queen Elizabeth I had proven to be a very successful political ruler. Under her leadership, the arts flourished, the economy boomed, and the military and naval forces protected English interests. She was the ultimate queen bee, in part because she never married and was thus able to manipulate suitors, countries, and policies to her favor.

Soon after Queen Elizabeth I died, her beekeeper, Charles Butler, published *The Feminine Monarchie* (1609). On the surface, the book reflected a dominant philosophy of seventeenth-century England—that is, nature was a model for human virtue. Butler wrote of the bees: "In their labour and order at home and abroad they are so admirable that they may be a pattern unto men both of the one and of the other."[3] The bees were loyal to the queen, refusing any type of anarchy or oligarchy. They labored incessantly for the good of the commonwealth. Therefore, according to historian Kevin Sharpe, "The keeping of bees was a pastime that was a lesson in statecraft and also one in personal conduct."[4] Sharpe's thesis works well with Frederick Prete's argument that although

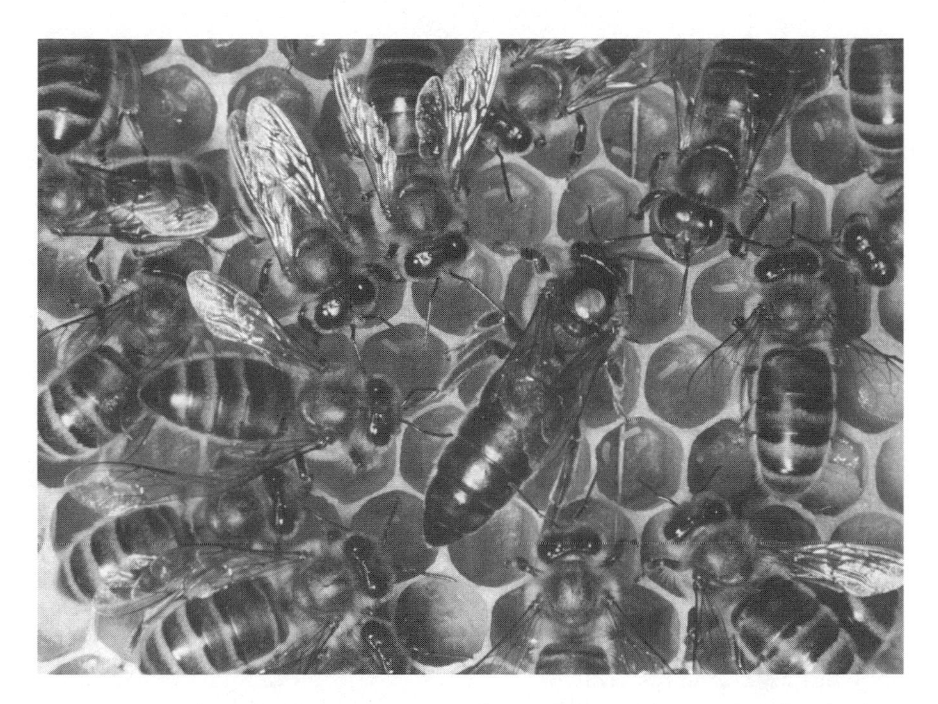

I.1. Queen and worker bees. Courtesy of *Bee Culture*. The queen is longer than the worker bees. Primarily an egg-laying machine, she is encircled by workers who groom her and transmit her pheromone to the rest of the hive. Pictured here are the house bees tending the queen. Charles Butler (1609) recognized that the monarch of the hive was a queen. Until then, people assumed that a king bee ruled the hive.

initially British bee books were used to teach women how to be better nurturers, Butler wrote to men and women, instructing them all in ways to be better members of the commonwealth.[5]

However, Butler's book had quieter and more indirect consequences. In very simple terms, he suggested a queen—not a king—was responsible for laying eggs in the hive. Thus, writers used the hive to reinforce hierarchical and patriarchal power structures. Butler challenged this cultural norm when he classified bees into three types: the queen, her female workers, and the male drones. His book had two strong implications: the first was that the queen "ruled" the hive, although we know that

I.2. Three types of bees. Courtesy of *Bee Culture*. Pictured here are three types of bees—a queen, a drone, and a worker. The queen is taller, the drone wider. The worker bee specializes in various tasks, depending on the needs of the hive. When seventeenth-century English thinkers transfered the drone label to their poor, the English began to colonize the New World with a value system based upon thrift, work, and stability.

such a phrase is quite misleading. A more serious, but unintended implication of Butler's research was that the unemployed poor were considered drones. In short, Butler reordered the hive for the English, and as such a brief introduction is in order.

We know now that bees aren't actually "governed" the way seventeenth-century English society wanted to believe colonies were. Instead, order within a hive is maintained by complex interactions among the queen, her female workers, and the male drones. Quite frankly, the queen is nothing more than an egg-laying machine. However, if she is strong, her presence in the hive sends signals to the rest of the workers that the colony is healthy. The workers clean the hive, guard the entrance, build cells, find pollen and nectar, store honey, raise brood, and even maintain temperature control through the changing seasons. The male drones have but one function: to mate with the queen. Because there is not enough room and food to feed the drones through the winter, the worker bees dispose of any drones remaining at the end of autumn.

I.3. Drone bee. Courtesy of *Bee Culture*. The drone is fatter than the other bees and has bigger eyes, which assist him in finding the queen when she goes on her nuptial flight. Other than mating with the queen, the drone does very little in the hive.

Given that the worldview at the time was to find lessons in nature, the image of lazy drones dying at the end of summer had powerful implications. The concept that labor was a virtue was readily adopted by the seventeenth century, because after all, the honey bees followed the natural and divine laws organizing a commonwealth. When Butler first hypothesized that the queen ruled the hive and that drones do nothing, he inadvertently provided a convenient analogy for seventeenth-century English writers, clerics, and politicians. These groups applied the drone label to thousands of unemployed and starving people.

King James and King Charles I, the monarchs who followed Queen Elizabeth I, ruled during unstable times, and their laws and ineffective social policies exacerbated the problems, leaving many poor people vulnerable. Even though England enjoyed economic prosperity during the seventeenth century, existence was hand to mouth in many regions. Peasant people were bound to the land in feudal arrangements or marginalized when the land was sold to yeoman farmers who did not adhere to the feudal system. "The old undeveloped agrarian society did not adjust with sufficient rapidity to provide employment for the thousands of laboring poor," explains historian Carl Bridenbaugh. Famine, late frosts, droughts, and damaged crops—the country suffered all of these during the early 1600s, and royal authorities refused to provide any financial or social relief.

Nor did the century get any easier. Bad weather and late frosts affected corn crops, accentuating the effects of an economic depression. In his diary, farmer Walter Yonge in Plymouth wrote about "three successive years, beginning in 1607, in which severe frosts or heavy rains caused an 'extreme dearth of corn.'"[6] Hard times plagued the English countryside from 1619 to 1624, from 1629 to 1631, and from 1637 to 1640. Because roads were so bad, many people could starve in one town while another town enjoyed great prosperity.

Ever fearful that the poor classes would rebel, especially in 1623 and 1630, the royal authorities divided the poor into two classes: the impotent poor and the idle. The royal courts perceived poverty to be "symptomatic of a deterioration of public order."[7] The impotent poor consisted of disadvantaged widows, the mentally and physically challenged, and orphans. Small amounts of relief for these people were available through church networks or feudal relationships. Moral judgment was not applied to this class. These people were considered unfortunate, but not immoral. The idle poor, on the other hand, included the unemployed, abandoned wives, unmarried mothers, beggars, and migrant workers. Royal authorities thought these people immoral because they did not work. The civil government feared that such unemployed masses might organize a rebellion.

Philosopher Francis Bacon best defined the fear of rebellion in his essay "An Advertisement Touching upon a Holy War," written during the famine year of 1622. Bacon proposed that masses of people be compared to various species of animals in the natural world, if left unchecked by the civil authorities. Historians Peter Linebaugh and Marcus Rediker explain why Bacon was so influential among the royal classes: "By taking his terms from natural history—a 'swarm' of bees, a 'shoal' of seals or whales, a 'rout' of wolves—and applying them to people, Bacon drew on his theory of monstrousness. These people had denigrated from the laws of nature and taken 'in their body and frame of estate a monstrosity.'"[8] Even though the reasons behind English poverty were complex, Bacon and writers before him simplified the issue by blaming civil unrest on the destructive capabilities of people, comparing them to a swarm and giving a negative connotation to honey bees that differed markedly from the Roman and Greek writers.

During the 1600s, therefore, the honey bee could be a double-sided mirror in English writings: it could signify order, or it could signify mass destruction. Bacon's essay best reflects the English mind-set regarding poor people, but many English writers found this drone image a convenient analogy to convince poor people to go to the newly emerging colonies in America. For instance, the image of honey bees was used in sermons, tracts, scientific proposals, and travel literature. Scholar Karen Ordahl Kupperman quotes John Cotton, who wrote: "Nature teacheth bees to doe so, when as the hive is too full, they seek abroade for new dwellings: So when the hive of the Common-wealth is so full, that Tradesmen cannot live one by another, but eate up one another, in this case, it is lawfull to remove."[9] Cotton was not the only one encouraging people to "hive off" to the New World.

Hiving off is a term synonymous with swarming. When colonies become too crowded, the queen will lay eggs for another queen, and then make preparations to leave the hive in search of a new place to live. Entomologist Tom Webster states that hiving off is when "one colony divides into two, i.e., hiving off is colony reproduction. Honey bees, like people, are highly social and do not thrive as isolated individuals. So, they must venture off in large groups when they are ready to establish themselves in new locations."[10]

And large groups were exactly what the seventeenth-century royal authorities were frightened of. More poor people seemed to be in the streets, even though women were waiting until later to get married. The transfer of the economy from large estates to small farms required fewer workers and more specialized techniques. When peasants were displaced, they often drifted to cities for lack of anyplace else to go. To those in authority, the poor had the potential to organize and rebel. Just as the drone imagery was convenient to apply to English beggars, so too the concept of swarming became convenient to apply to the poor masses.

Thus, politicians, clerics, and entrepreneurs adopted a biological model to justify a social phenomenon. Writers such as John Cotton, Richard Hakluyt, Richard Eburne, and Francis Bacon were convinced that overcrowded conditions, lack of jobs, and poverty were valid reasons for the mother country to hive off to the New World. To quote Kupperman, "Just as bees swarmed from the overfull hive, English men and women should leave England, groaning under its heavy burden of overpopulation, for the good of the commonwealth."[11] From an average English person's perspective, the New World was the perfect place for idle people to swarm: plenty of room, plenty of work, and plenty of exploitable resources, including poor men.

The English also wanted to extend philosophies from translated Greek and Roman beekeeping and agricultural texts. In fact, the Greco-Roman myth of Aristaeus prefigured the biblical promise of a land of milk and honey. In ancient Greece, the story goes, honey bees lived in rocks and caves, but Aristaeus managed to domesticate bees, thus ensuring an adequate diet for the Olympian gods and goddesses. Honey was needed for celestial nectar and ambrosia (the food and drink of the gods). Celestial nectar was made from fermented honey and water; those deities who drank it returned to health. Ambrosia was made with milk and honey; those deities who ate it enjoyed eternal beauty and bloom. Aristaeus supplied the Olympian deities with their food and drink and lived a carefree shepherd's life—until he saw Eurydice. Eurydice was the intersection point between two myths—one involving the bee-loving Aristaeus, and the other involving the talented musician Orpheus.

The Orpheus myth remains well known even in the twenty-first century: Orpheus could play so well that he charmed wild beasts. When he and Eurydice married, everyone assumed they would live happily ever. However, while Eurydice was walking in the fields after the wedding, Aristaeus fell in love with her at first sight and began to chase her. While she was running from Aristaeus, Eurydice stepped on a viper. She died from its bite. She was taken to Hades, where Orpheus descended to try to bring his wife back. But Greek gods were not known for their forgiving ways. To punish Aristaeus, Olympian nymphs smashed his hives and killed his bees. And for most people, that was the end of the myth.

But the myth continues for those of us who love bees. Broken-hearted, Aristaeus appealed to his mother, Cyrene, the sea goddess. Cyrene taught Aristaeus how to negotiate with Proteus, the shape-changing god who took care of the sea calves. When Proteus arrived at midday with his sea calves, he set them loose to graze and then took a nap. Aristaeus tied him in fetters, so when Proteus awakened, he was trapped. Proteus tried to change shapes, but Aristaeus had tricked him. So Proteus worked out a deal with Aristaeus that would atone the death of Eurydice and also let the beekeeper have his bees. In the myth, Proteus tells Aristaeus, "Select four bulls of beauteous form and as many heifers whose necks were yet untouched by yoke. Sacrifice these animals on the altars. After their throats have emitted the sacred blood, leave their bodies in a leafy grove. Return in nine days and see what will befall." Aristaeus returned to find the carcasses of the cows teeming with swarms of bees. This image of cows and bees—so symbiotically connected—embodies the pastoral traditions in Greece and Rome. In other words, once this myth establishes a relationship between bees and cattle, it verifies an agrarian and interdependent civilization promoted as an ideal throughout the centuries.

The Aristaeus myth established this popular misconception about the symbiotic relationship between bees and cattle, but it was not the only cultural document to perpetuate the metaphor of a land where milk and honey are interdependent. Translations of Pliny's *Natural History* (available in 1601) and Virgil's *Aeneid* and *Georgics* impressed on the English mind the importance of bees in those ancient worlds. Virgil detailed how to produce bees from oxen. Pliny advised his readers how to rejuvenate dead bees: by burying them in a carcass of an ox in dung. The linking of these images in mythology and literature reflect these societies' perception of the symbiotic relationship between cows and bees. The myth emphasized to farmers in the Greek and Roman societies that their lands needed to be kept fertile and productive.

This myth and the texts that sprang from it provided a convenient analogy for the ideal place, a civilization where people could go for a new start in life, but one that could also extend the prosperity from better times. Although the seventeenth century was notorious for famines and overpopulation, English writers conveniently drew parallels from prosperous Greek and Roman societies to their own. The English saw themselves as being direct, linear extensions of superior Greek and Roman traditions; King Arthur was widely regarded as the descendant of Aeneas, and Elizabeth I could be seen as the English version of Dido. It was only natural that England, like Rome, should extend her power via swarms of people to the New World. William Symonds (1609) wrote of an England "where [the] mightier like old strong bees thrust the weaker, as younger, out of their hives."[12] Significantly, the English writers secularized the image of the honey bee when making the parallels between the two time periods. It was no longer a direct link between human and Olympian society. In an effort to use a symbol that would justify colonization, some English writers stripped the honey bee of its status as a divine symbol, which had been unquestioned in Greco-Roman society.

England did thrust the "younger" bees out of the hive, expecting the "drones" to go to the New World and immediately reap the benefits of farming, as if the New World were like England. However, until Samuel Hartlib finally published the first practical book for English farmers and beekeepers in 1655, the first English colonists had very little training in apiculture. Hartlib's *A Reformed Commonwealth of Bees* marked a new era in English agriculture because Hartlib wrote about English lands and plants, not Greek ones. According to Timothy Raylor, the book was the "high point" in Hartlib's agricultural reform efforts because it was not based on "classical authority or native experience" but rather verifiable experiments and analyses.[13] Hartlib's book focused on new advances in English beehives instead of Greek beehives or practices because he wanted to persuade Parliament to quit relying on the slave trade and colonies for sugar. He was convinced that all of England's sweetener needs could be produced by English beekeepers, if there were just more beekeepers and more money to pay them. If the financial resources were kept within England, Hartlib reasoned, the country would not suffer such economic hardships.

Hartlib's original intent had been to convince Parliament that England should be a nation of self-sufficient beekeepers, but he was very interested in the beekeeping possibilities in the New World. He sent a copy of *A Reformed Commonwealth of Bees* to John Winthrop, the first governor of Massachusetts. Hartlib's expectations for English beekeepers did not succeed, but Hartlib "introduced a high level of co-operation and institutionalism to the search for new apicultural techniques," according to Raylor.[14]

Because the New World was still undeveloped through much of the seventeenth century, the colonists were not in a position to capitalize on Hartlib's new beehive designs. Furthermore, the slave trade involved too much money for American and English businessmen to discontinue their involvement. The slave trade and sugar became inextricably linked by the seventeenth century. Once cheap sugar became readily available, the English and Americans did not really care *how* their sugar was processed and ignored the slavery question for two centuries. Thus, Hartlib's plans for self-sufficiency, at least in terms of sweeteners, were never taken seriously.

But we are getting ahead of ourselves.

The immediate crisis that forced the issue of New World colonization was not sugar, at least in the early 1600s. The crisis developed from a combination of dire circumstances happening at the same time: bad weather, lack of food, and lack of work. The obvious answer to relieve overpopulation was the New World. America became the place for those displaced or dispossessed: "The drones of England became the workers of America," explains Kupperman.[15] New England offered poor people a chance to escape land contracts, the vagaries of bad weather leading to bad harvests, and increasing taxes on very little yield. But the popular analogy between the New World and the hive simplified the complexities of poverty. And the comparisons of males to drones in travel literature did nothing to solve outdated land or social policies.

The English desperately wanted social and political order, but changes were happening too rapidly on too many fronts—social, economic, agricultural, scientific, political. When the English began to colonize the New World, *their* New World, with their own marginalized people, they adopted the metaphor of the hive—along with its ancient classical traditions and simple gendered social structure—to categorize people and behaviors. Many writers and colonists liked to project an image of America as an orderly, ideal hive, even though early America was a frontier when compared to Europe's standards of society.[16]

Furthermore, early colonists were determined to hold onto the rhetoric of apiculture, despite how often the ideal beehive model of society failed to meet reality in Massachusetts and Virginia: "Endorsed by both ancient wisdom and nature, the hive seemed to offer a perfect model for colonization," writes Kupperman.[17] The keyword in Kupperman's sentence is *seemed,* for the hive model was a beautiful theory that did not withstand the harsh realities of New World life.

Just as the English craved order, so too did the colonists, and the beehive image represented efficiency, industry, and, most important, social stability. The associations with this image became much more powerful than colonial beekeeping was. The negative associations colonists had of idle people who did not or could not work transferred from England to the colonies. Leaders consistently equated "idle poor" with "drones" in Jamestown and the Massachusetts Bay Colony. In other words, the colonists were preconditioned to see people who weren't "busy as bees" as lazy and slothful and deserving of the fate that happens to drones at the end of summer: to be forced out of society.

Four centuries after hiving off from England and Europe, Americans know a thing or two about bees. We now know that the larvae of flies, not bees, nest in the carcasses of cattle. It turns out the beautiful myth of Aristaeus, Cyrene, and Proteus is reduced to plain old larvae. Virgil and his peers erroneously thought the larvae of the fly *Eristalis tenax* or those of blowflies were the larvae of honey bees because they look very similar.[18]

We also know how dangerous it is to simplify society by the use of examples in nature. However, many Americans still value the honey bee as a symbol of thrift and industry. This value seems to be one of the lingering philosophies from seventeenth-century England, in which the royal authorities and clergy dictated that the lower classes and unemployed should be "busy as bees" so they would not rebel. When the English began to label their own members of society as "drones," they privileged a new set of values based on work, thrift, and efficiency. The American Dream still seems to be based on these very values. And if somehow people do not attain the American Dream, we tend to think that they have not worked hard enough or did not save their money—in short, they are too much like drones. It could be argued that many American social policies—so conscious of work, labor, and time—are still based on the beehive model first adopted during the seventeenth century in

England. For all its rhetoric of new opportunities, America still sees poverty as a sin, as if somehow the poor aren't thrifty or busy as bees.

This book is about how the honey bee has been perceived in America and how those perceptions have changed as the country developed through the centuries. Although the honey bee retains its religious importance in some American communities, it gradually has taken on secular values as America became industrialized. Thus, the book is arranged chronologically, for we trace the bee's secularization and westward migration together. Toward the end of the twentieth century, the honey bee suggested even more ideas, such as freedom, political dissonance, and domestic dissatisfaction, as marginalized communities finally found expression in music, literature, and film. Furthermore, because there are many types of bees in North America, including solitary varieties, I will use the term "honey bee" to distinguish this insect from other, less social bees.

Chances are, if you've read this far, you value esoteric information, and so you may enjoy thinking about subtle ways in which bees impact the American society. Keep reading. You may find that the simplest symbols that grace our banks, sidewalks, and billboards are far more complicated than you had ever imagined. You may hum a blues or gospel tune with a little more understanding of the complex sociological conditions from which the singer emerged. You may rethink your standing on social policies. You may even find yourself buying local honey and supporting regional beekeepers. The truth is, my goal has been much simpler: to provide enough evidence to serve as a valid response to Karl von Frisch's question about the need for bee books. If I have provided enough evidence to convince you that there is always room for one more bee book, then this worker bee may rest.

Chapter 1

Bees and New World Colonialism

> If the Lord delights in us, then He will bring us into this land . . . a land which flows with milk and honey.
>
> —*Numbers 14:8*

European settlers often quoted this biblical phrase to justify their colonization efforts. As long as settlers had cattle and bees, they could be assured of the basic essentials—food, wax, medicine, candles, and clothing. So powerful was the Bible verse that even though cattle and honey bees did not exist in North America, colonizers envisioned the New World as having them in the immediate future. But each European country handled its land acquisition in different ways. Whereas French and Spanish explorers conducted formal negotiations with Native Americans for land rights in the New World, the English settlers merely appropriated land that appeared unused. Summing up the English mind-set that land should be cultivated or else it was wasted, George Peckham states, "I doo verily think that God did create lande, to that end that it shold by Culture and husbandrie, yeeld things necessary for mans lyfe."[1] Because English settlers did not recognize or acknowledge Native American agricultural patterns already in place, they fundamentally changed the landscape by bringing cows, bees, apple trees, and even, inadvertently, mice.[2] But the successful migration of bees, cattle, and the colonists themselves was not a foregone conclusion. In fact, the first attempts by English colonists to send bees to North America ended up in the Bermudas when a storm blew the *Sea Venture* off course in 1609.[3] America could become a land of milk and honey, if bees and cows could survive the transatlantic journey and if the colonists themselves could survive the illnesses, difficult weather, and Native Americans.

The first two English settlements—Jamestown and Plymouth—were not chosen wisely. Jamestown was settled in the middle of powerful Algonquian chief Powhatan's domain. And the *Mayflower,* the ship carrying the Pilgrims, was blown off course from its initial destination of Virginia to the cold, rocky Massachusetts coast. So even though the English had been going to the colonies since 1609, neither place was suitable for bees until 1621. The reality of frontier America hardly lived up to the biblical promises of plenty assured by those back home, some of whom had never seen the New World.

Colonists in both Jamestown and Plymouth died at an alarming rate from a number of different causes. At the end of Jamestown's first year, 1607 to 1608, only 38 of the original 108 settlers were alive. "The winter of 1609 was the notorious starving time in Virginia," Kupperman writes, "when the population dropped from 500 to sixty in six months."[4] Similarly, almost 50 percent of the Pilgrims died during their first year in Massachusetts.

Leaders used the beehive metaphor to rally colonists' spirits. Depression was not recognized as a sickness during the seventeenth century. But many Jamestown colonists, separated from England and suffering from malnutrition, were sick and indifferent. Furthermore, the Jamestown colony was made up only of men. Rather than being treated with sympathy, they were labeled as lazy or "idle drones." Thinking that the cause was primarily a lack of leadership, Robert Gray wrote in 1609, "The Magistrate must correct with al sharpenesse of discipline those unthriftie and unprofitable Drones, which live idly."[5] Lord de la Warr, sent to Virginia in 1610 to provide leadership, managed to excuse himself from being labeled a "drone" in his summation of scurvy, but he admits that "which though in others it be a sicknesse of slothfulnesse, yet was in me an effect of weaknesse, which never left me, till I was upon the point to leave the world."[6]

Oblivious to titles and ranks, Captain John Smith often used the term *drone* to refer to all colonists, correctly realizing that men needed structure and activity if they were going to survive. Smith forced the early colonists to revise their new roles in the New World. Immediate goals helped dispel the gloom and low morale that resulted from lack of good, clear leadership and disappointment with the reality of frontier life. Believing that regular activities would keep his men healthy, Smith campaigned for the men to build houses and forts, clear land, take care of the sick, and garden.

By 1621, the Virginia Company was sending ships loaded with "divers sorte of seed, and fruit trees, as also Pidgeons, connies [rabbits], Peacock maistives [mastiffs], and Beehives," according to an invoice sent from the Council of the Virginia Company in London to the Governor and Council of Virginia.[7] Clearly, the Virginia Company and Captain Smith recognized the importance of keeping their men active and not letting lethargy set in because, although depression was not recognized as a disease, *idleness* was the most-used term, which suggested moral failings. "Just as in the earlier Jamestown mortality," according to Kupperman, "there was the conviction that the 'mother and Cause' of the contagious disease was 'ill example of Idleness.'"[8]

Whereas Jamestown was founded for commercial reasons, Plymouth was founded for religious reasons. The Pilgrims were a Protestant group convinced that the Church of England was impossibly corrupt. They were one small fraction of a larger group called the Puritans, who believed that the Church of England should be the official church, only "purified." However, many Pilgrims finally came to believe, according to Kupperman, that "only cutting the colony off from supervision from England could allow it to begin to create an improved version of English society."[9] Thus the Pilgrims and Puritans differed in terms of their decision to stay in England. The Puritans chose to stay in England longer. But in order to practice their form of Protestantism, the Pilgrims left England. After waiting in Holland for years, the Pilgrims were even more determined to succeed in North America. The Pilgrims set sail on the *Mayflower* bound for Virginia, but as stated earlier, the ship was blown off course and landed at Plymouth Rock.

Forming a joint-stock company with London Merchants, a company willing to finance the Pilgrims' venture to the New World, the Pilgrims themselves represented a share of stock. "The colonists were to work for the company for seven years, at the end of which they were to receive title to the land and all other assets were to be liquidated in accordance with stock ownership," explains William Sachs.[10] In contrast to Jamestown, whose inhabitants did not own shares, the Pilgrims were the investment in the New World. Later, when the Puritans organized the Massachusetts Bay Colony, top management leaders such as John Winthrop decided to relocate to America. Absentee management was not the problem at Plymouth or Massachusetts Bay Colony that it had been in Jamestown.

Another major difference separated the Massachusetts Bay Colony from Jamestown and Plymouth. The Massachusetts Bay Colony sent more than a thousand colonists and thus ensured that the colony would have a good reserve of morale, workers, and adequate resources from which to begin. In this way, the English clergyman Richard Eburne writing in 1624 was correct in comparing colonists to bees: "The smallest swarms do seldom prosper, but the greatest never lightly fail."[11]

Because the Plymouth and Massachusetts Bay colonists brought their families with them, the psychological effects of colonization seem to have been tempered earlier than in Jamestown, an issue that Kupperman addresses: "The key to getting people to work together on common goals, paradoxically, was to allow them to work for themselves and their families. . . . Families, made possible by the emigration of women, rendered individual efforts meaningful; passing an estate on to one's children was the goal of hard work and deferred gratification."[12] This combination of theological, financial, and familial interests was never realized by other commercial ventures interested only in how much money could be made in America. In fact, when it was theorized that the colonists would be susceptible to Indian attacks, colonist William Wood refuted that notion. Wood realized that colonists who have property to protect would band together: "When Bees have Honie in their Hives, they will have stings in their tailes."[13]

Massachusetts also had a number of different economic options besides tobacco, including timber, fishing, and shipping. Sachs estimates that from 1630 to 1640, an estimated 20,000 immigrants arrived in

Boston to carry on the work of building ships, fishing the Atlantic, and finding timber for ship masts and naval yards.[14] When trade opened with the West Indies and southern Europe, business boomed in Boston. Yet even though the strong Pilgrim and Puritan religious communities tempered the emotional effects of colonization, the same fear of idleness pervaded Puritan and Pilgrim writings. Massachusetts governor John Winthrop sadly wrote to his wife, "I thinke here are some persons who never shewed so much wickednesse in England as they have doon here."[15]

Not all were wicked. In fact, because most colonial women married, the term *good wife* came into existence, around which developed a code of ethics that governed female life in northern New England from 1650 to 1750. Good wives had legal rights in colonial America and had more freedom than nineteenth-century women had. They often shouldered the responsibilities of farms and shops when men were away at sea or on trips. *Deputy husbands* was another term given to wives who learned their husbands' trades. "Since most productive work was based within the family," writes Laurel Thatcher Ulrich, "there were many opportunities to 'double the files of her [a wife's] diligence.' A weaver's wife . . . might wind quills. A merchant's wife . . . might keep shop. A farmer's wife . . . might plant corn."[16] It could be just as likely that a baker's wife would visit a hive if honey were needed to bake, by extension. Certainly, according to scholar David Freeman Hawke, the colonist's wife "still performed a number of fixed duties as in England—prepared meals, milked cows, washed clothes, tended to the kitchen garden and the beehives if the family had them. But her role was no longer limited to household chores . . . she now became the husband's partner in the fields."[17] When the English colonies merged with New Amsterdam in 1664, the term *good wife* took on even more possibilities because Dutch women had more freedom than English women, and their inventories of kitchen gardens mentioned beehives.

Settlers cherished the image of the beehive for the stability and order that they couldn't enjoy in their home countries, and when they began to prosper, they "hived off" to form new communities.[18] In Massachusetts, colonists created structured, organized societies based on English township patterns: "In the middle was the common, with the main streets running around this rectangle. Nearby the townsmen built their church, adjoining the minister's house and church school. Each settler was given a home lot, which varied from a quarter of an acre to twenty acres, where he might plant an orchard and put in a small garden. Not far outside the village each colonist was assigned a strip of land for cultivation."[19] This arrangement made it easy for English communities to plan communal agricultural activities, such as harvesting, planting, and sowing.

In order to hive off in the North American colonies, village leaders followed a process similar to England. A community petitioned the general court in Massachusetts for permission to send settlers to a new site: "Those designated to migrate would, on the appointed day, set forth to the new location under the directions of a few leading citizens authorized to supervise the journey. On arrival at the specified site, each family was again allocated a home lot, a strip of arable land, and the privileges of the common."[20] Because honey was an important commodity, many towns included plans for an apiary after orchards were planted. An Englishman named William Blackstone immigrated in 1623, bringing a bag of apple seeds with him. He planted these seeds on Beacon Hill, Boston, but they did very poorly. The Puritans remembered that the orchards in Europe had honey bees, and thus, they requested that bees be sent. The bees soon arrived and thrived. So did the apple trees. When Blackstone moved to Rhode Island, he took his seeds with him.[21]

Danvers, Massachusetts, established a town apiary in 1640. Danvers was started in 1636 after a grant of 300 acres was assigned to John Endicott of Dorchester, England. He called his land Orchard Farm, but eventually, when other families were established near Salem, the two towns merged interests. This community apparently had a thriving beekeeping population: "In 1660, a stand of bees was appraised at five pounds in this town. . . . An inventory of T. Barnard's estate, made after he was killed by Indians in 1677, included eight hives of bees." The town of Newbury, Massachusetts, was fortunate to hire its own beekeeper, John Eales. In 1645, he was hired to follow "his trade in beehive making."[22] Eales would have made straw skep hives, which were popular during the seventeenth century.

Although it is considered old-fashioned by contemporary standards, the skep hive signaled a remarkable advance in beekeeping. For centuries, Egyptians kept their bees in mud tubes, the Romans and Greeks kept bees in clay pots, and many cultures had log hives, especially those

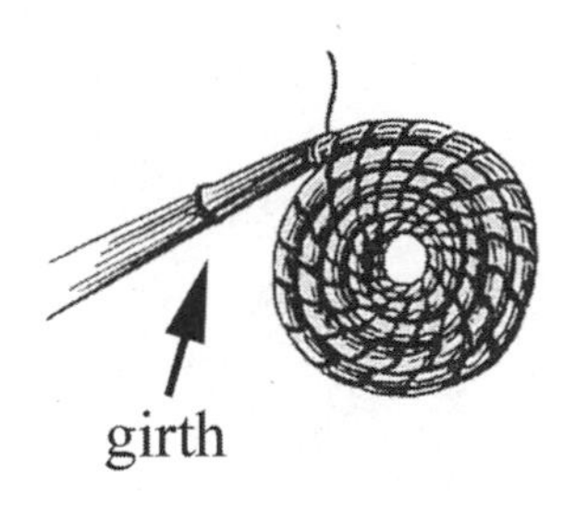

Right, 1.1. Skeps, Girth. How a skep is started. Originally illustrated in Herrod-Hempsell, 1930. Courtesy of Kritsky, April 2003. Making a skep was an art and a skill. The girth was the first step in the process. Many villages hired skeppists to meet the beekeeping needs of the community. John Eales was one of the first skeppists hired in colonial America. Given the unsettled economy during that time, he eventually died a pauper.

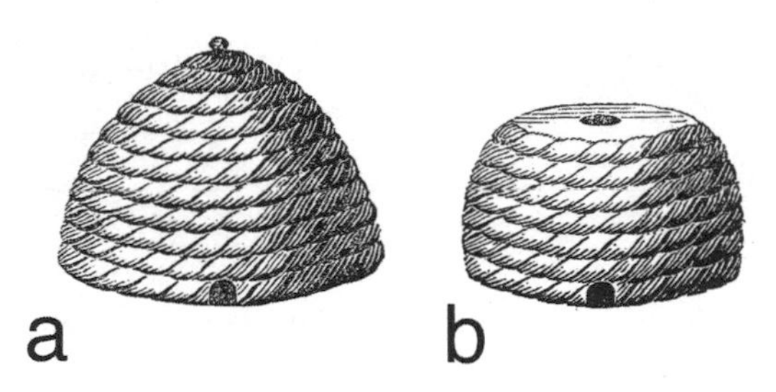

Above, 1.2. Skeps. Typical bell-shaped skep hive with handle. Originally illustrated in Bagster (1838). Courtesy of Kritsky, May 2003. Skeps derived from the Scandinavian word for "pot," which the straw hives resembled. But straw was more plentiful and easier to carry, especially when skeppists added handles.

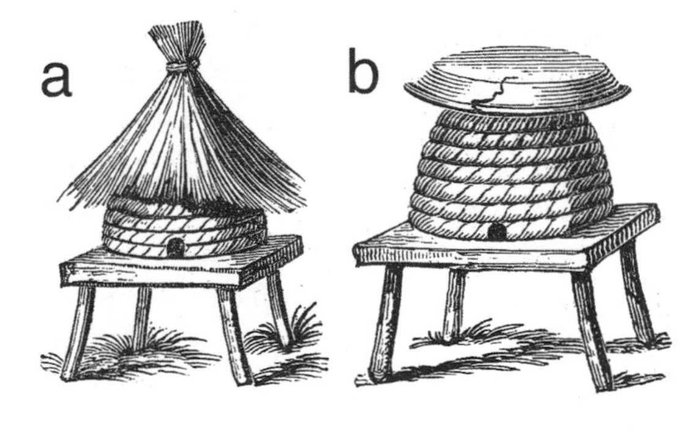

Left, 1.3. A thatch hackle tied over a skep hive; a milk pan used as a cover on a flat skep. Originally illustrated in Bagster (1838). Courtesy of Kritsky, May 2003. These illustrations indicate the versatility that beekeepers had with skeps, which represented a new technological advance in beekeeping.

1.4. A skep beekeeper placing his hive on an eke. Originally illustrated in Bagster (1838). Courtesy of Kritsky, May 2003. An eke extends the hive, which provides more storage for a full hive of bees and discourages swarming.

in Western Europe. In fact, the word *skep* comes from the Scandinavian word for *pot.* But as people settled the treeless plains in eastern England, they developed a straw hive that could take the place of heavy log hives. According to historian Gene Kritsky, "The skep hive relied on easily obtained raw materials for hive construction and made movement of colonies a much easier process. This hive would enable beekeepers to increase the number of colonies that they managed and provide an easier method of swarm capture."[23] Skeps were lightweight and could be easily repaired with materials at hand. Bee skeps would have been in demand in the New World colonies just as they were in England. "Beekeepers needed to have two skeps for every colony: one skep for the colony and an empty one for its swarm," explains Kritsky. "Thus, there was a demand for good skeps to replace ones lost during the previous year, or to increase the number of hives."[24] In fact, an anonymous saying from an *American Bee Journal* column indicates that beekeeping was a popular activity in the colonies:

"There is scarcely a village in the country that might not readily keep as many hives of bees as there are dwelling houses in it."[25]

Some Pilgrims in Plymouth decided to settle in Connecticut. Although these pilgrims brought their hives, they found winters in Connecticut to be as difficult as they were at Plymouth. Furthermore, the Pequot Indians were more protective of their land, so the threat of attack was a constant worry.

Colonists in other regions did not hive off in as organized a fashion as in Massachusetts. For example, Cecil Calvert, whose father was a member of the London Company, obtained a charter in 1632 for 10 million acres in Maryland. Although there were many English settlers, the region quickly absorbed a number of French Huguenots. In 1685, French Huguenots fled the Edict of Nantes and settled in Maryland and Pennsylvania. Although the French had maintained a consistent presence in the New World since the 1500s, they primarily had been to the west and north of the English colonies. The Huguenots who arrived on the East Coast brought sophisticated gardening and beekeeping skills.[26] An old Clinton County bee master named Tommy Simcox recorded the following: "Early Huguenot [beekeepers] in the West Branch Valley [Pennsylvania] were wont to start upstream with their bees on flatboats every spring, traveling by night so as not to excite the bees and anchoring by flowery glades by day, each day giving their swarms a new and delicious pasturage. By the time they reached, 'Les Fourchetts,' or the Forks in what is now Keating, the weight of the honey lowered the boat to the water's edge and the return journey began."[27] Simcox even provided the names of early Huguenot beekeepers in the region: Old Adam Kres, Jacob Carnes, Michel Hulin, Moise Godeshal, and Francos Bastress. These Huguenot beekeepers capitalized on the easy availability of bee trees in the region before William Penn organized Pennsylvania.

Although the English were the most visible colonists, they were not the only ones settling on the Atlantic seaboard. In many ways, the Swedes were acclimated to the Pennsylvania colonial frontier.[28] After Sweden passed the Forest Law of 1647, many Finnish foresters, who had been encouraged to harvest Swedish forests, were punished and imprisoned. One way out of prison was to go to America.[29] The Finno-Swedes were "forest pioneers" and quickly applied their forestry methods in the New World. Historian Frederick Hahman reports that the Finno-Swedes brought German bees to the colonies as early as 1638.[30] Pennsylvania offered plenty of white pine and hemlock forests, especially west of the Susquehanna River.[31] In fact, according to Terry Jordan and Matti Kaups, "Indians complained to the Dutch that the New Sweden settler builds, plants on our lands without buying them or asking us."[32] The Indians also tended to associate the black bees with the bullets used in rifles, and thus remained suspicious of the honey bees for a long time, according to Hahman.

Honey was very important to the Swedes, however, especially as the settlers extended into the Midland. They needed the quick, easy energy that honey supplied. Evidence suggests that the bees did very well in colonial Pennsylvania. Although the Swedish crops often failed, "Bees thrive and multiply exceedingly," wrote a newly arrived Welshman; "The Sweeds often get great store of them in the Woods, where they are free for any Body."[33] Jordan and Kaups suggest that the Swedish colony succeeded in ways that the English did not, for the Swedish colonized the Pennsylvania forests with the least possible risk.

The Dutch, who were more prepared than the Swedish colonists, first arrived in 1624 to establish New Netherlands.[34] When first sending out an exploratory group, the Dutch East India Company considered bee skeps essential equipment. Later, when the West India Company was granted a charter to settle New Netherlands, it used flowery descriptions promising a New Canaan to entice the Dutch to leave the comfort and security of Holland. Unlike England, Holland was quite tolerant of religious differences, and many people were content to stay in Holland. The Pilgrims stayed for ten years, for instance, before sailing to Plymouth. In order to populate the New Amsterdam region, West India Company writers produced pamphlets in which, to quote historian Oliver Rink, "hyperbole knew few limits."[35] In fact, he quotes a Dutch writer who claimed that New Netherlands was "a maritime empire where milk and honey flowed."

New Netherlands was far from being a new Eden. In addition to the normal frontier challenges and hardships suffered by the English colonies, New Netherlands colonists could be intolerant of those who were not members of the Dutch Reformed Church. Blacks were imported as slaves, and Jews were treated poorly, although they did stay and eventually bought property. Quakers and Lutherans also were not welcomed.

Only after official censure from authorities in Holland did Governor Peter Stuyvesant relax his control of the colony.

But the colony did have honey. These settlers often kept bee skeps on the south side of their homes.[36] One anonymous source quoted by Eva Crane stated, "You shall scarce see a house, but the South side is begirt with Hives of Bees, which increases after an incredible manner."[37] The south side of the house generally had a porch that extended the length of the house, and thus the bees would have been sheltered from the elements. Because the architecture suggests that bees were in close proximity to the houses, Dutch women probably kept bees just as they tended to the kitchen garden. If they didn't have beehives, they might have had a Dutch oven, which was often called a beehive, for it "projected through the wall to the outside of the dwelling," according to Firth Haring Fabend.[38] When the English took over the New Netherlands in 1664, beekeeping was already established in this colony. It was a peaceful takeover because there had been a great deal of cultural interchange between the New Netherlands, Massachusetts, and Connecticut beforehand.

Upon his father's death in 1670, William Penn exchanged a debt of 16,000 pounds to King Charles II for 47 million acres of land in the New World. In 1681, a charter was granted, and he printed a pamphlet, "Some Brief Information on the Province or Region Called Pennsylvania Located in America." He advertised land to colonists in three languages—Dutch, German, and French. His efforts paid off. In 1683, many German Protestant communities came to Pennsylvania, New York, and Delaware. Exhausted by the Thirty Years' War and religious persecution, many groups such as the Mennonites, Hutterites, and Quakers found religious freedom in Pennsylvania. The Lutherans and the Dutch Reformed Church came for economic reasons. These groups, arriving from all parts of Germany, were proficient beekeepers and honey hunters.

A Mennonite Quaker from the Rhine Valley named Francis Daniel Pastorius crossed to Pennsylvania in 1683 aboard a "Noah's Ark of religious faiths—Roman Catholics, Lutherans, Calvinists, Anabaptists, and Quakers."[39] Pastorius was not a beekeeper, but he carefully recorded all that he read in a bibliography called the "Bee Hive." It contains approximately two thousand individual entries that Pastorius called "honey combs," according to scholar Alfred Brophy. Pastorius was widely read and wanted his sons to be as well. Although the Bee Hive was intended for his family, it was and continues to represent the most comprehensive collection of Quaker and non-Quaker publications available in the colonies during the seventeenth century. Furthermore, according to Brophy, the Bee Hive served as a repository for the Quaker experience in England. By identifying a wide variety of Quaker writings in early Pennsylvania, Brophy argues that Pastorius "provided a substitute for memories of life in England" in his Bee Hive bibliography.[40]

Further south of Pennsylvania, another experiment with colonization was happening. In 1670, the Virginia governor William Berkley, along with seven other investors including a Barbados planter named John Colleton, began a colony of rice and indigo plantations. As a colony, South Carolina offered cheap land. Slavery was considered the norm. Some prisoners settled in this region once they were released from prison in England. Whereas the beehive metaphor was constantly in the literature of sermons, diaries, and daily language in the New England colonies, it was nonexistent in the southern colonies, which tended to be founded later and by individuals rather than religious groups.

Contrary to the image of the Puritan work ethic, New World colonists did have fun. Honey hunting became one of the first American pastimes. Forests in Sweden, England, and Germany were tightly controlled by the state, but colonists were free to find bee trees in America. Bees did extremely well in the New World forests, providing plenty of opportunities for hunting. Furthermore, colonists even appropriated the word *bee* to refer to a social gathering. Bee historian Rollin Moseley explains the evolution of the term: "Settlers along the Atlantic Coast were delighted to find wild bees quite plentiful. Frontiersmen soon noticed that the tiny bees always worked in groups and began to call any social gathering that combined work and pleasure a 'bee.'" When communities organized new churches, social bees were held in order to compensate the preacher for his services: "All members of the community, whether they attended church or not, were solicited for gifts or work, clothing or food commodities," according to Moseley.[41]

Scholar Helen Adams Amerman suggests that the phrase "quilting bee" may have even originated in New Netherlands in 1660 when Dutch women gathered for the first social occasion of the season. Certainly, more preparation was taken by the hostess to make the affair fun and

1.5. Colonial husking bee, courtesy of the Library of Congress. No date recorded. H.W. Pierce drew this quilting bee, which was a festive occasion for colonial women. Men enjoyed "bees" to finish barns, harvest crops, or raise houses.

festive: "The [participants] could wear their Sunday go-to-meeting clothes, an excuse they seldom had, and it was more elegant than corn-huskings or apple-peeling parties. The women sent out invitations. . . . This was one of the rare times that the 'fore-room' was used."[42] Although the term *bee* does not appear in newspapers until 1806, in Washington Irving's *Knickerbocker's History of New York,* he credited the social gatherings to Dutch Governor Peter Stuyvesant, who served as a director for the West India Company when the earlier director Willem Kieft proved to be an ineffective leader for the new colony. Distributing fiddles to the black musicians, Stuyvesant called for "joyous gatherings of the two sexes to dance and make merry. Now were instituted 'quilting bees,' and 'husking bees,' and other rural assemblages, where, under the inspiring influence of the fiddle, toil was enlivened by gayety and followed up by the dance."[43] When people wanted to quench their thirst, they drank mead, a honey-based drink. Thus the bee affected leisure as well as culinary and linguistic patterns during the early colonial years, even though as an organized, systematic hobby, American beekeeping languished behind other countries.

Labor was a key issue in colonial America. If too many people were "idle drones," then the business of colonizing (that is, the planting of) the New World would not get done. Although the English initially relied on indentured servants, many landowners quickly turned to the markets selling African slaves to resolve the labor issue, especially in the South. Recognizing that tobacco and rice would make the colonies economically viable, Virginia and South Carolina legalized slavery first.

Dutch traders sold the first slave to Jamestown leaders in 1619. From then on, slave transactions were common between the Dutch and English colonists, until King Charles II ordered the English to take over New Netherlands in 1664 and added New York to his colonial properties. This takeover streamlined the slavery process. In 1698, the English Parliament ended the Royal Africa Company's monopoly on the slave trade. That year, the number of African slaves brought into North America jumped from 5,000 to 45,000. Slaves from Angola, Senegal, Cameroon, and other bee-friendly climates were brought to various ports in the New World, such as South Carolina, New Netherlands, Massachusetts, New Orleans, and Virginia. Many Africans came from honey-producing regions in Africa, especially Angola.[44] So many Angolan slaves arrived to his 8,000-acre plantation in Louisiana that Pierre LeMoyne Sieur d'Iberville named his plantation Angola in 1699.

During the seventeenth century, Africa supplied the world's largest amount of beeswax, although its value was ignored initially in tribal economies. Ironically, before the Portuguese arrived, many Africans threw beeswax away. But when Portugal conquered large areas in Africa to provide slaves for their sugar plantations in Brazil, the Africans adopted the word for candle for beeswax, *kandir.*[45] The Portuguese needed beeswax for two main reasons: church candles and lost-wax molds, which were used to create religious objects and weapons.

Lost-wax molds (*cire-perdue*) were used by many cultures during this time. In fact, Eva Crane divides her discussion of lost-wax casting among four civilizations: Mediterranean peoples dating from 3500 and 3000 B.C.; the Asian peoples from 2500 and 1700 B.C. (and it is still being used); the European cultures from 300 B.C. to the Renaissance; and the African peoples beginning in A.D. 800 to the 1700s. Crane describes the lost-wax mold process as follows: "A model was first sculpted in beeswax, and coated with pliable clay. . . . This was hardened by drying in the sun or by some other means. Next the whole was heated so that the wax of the model melted and was drained out or 'lost' through one or more vents. . . . To make the casting, molten metal—usually copper, bronze, brass, or gold—was poured in through an opening at the top of the mould and allowed to solidify, after which the mould was chipped away and discarded."[46] Crane further explains that large castings were done with lost-wax molds, and that ancient statues with complex intricate loops may be indicative of lost-wax molds. But this artistic skill died when slaves were brought to America. Although Crane does not speculate why, she does state the lost-wax mold process demanded a constant supply of beeswax. Such a supply was not readily found in the colonies.

Although West Goree was the main export center for slaves and beeswax in Africa, beeswax came from all over West Africa, suggesting that the people from these regions were very skilled bee hunters, although Carl Seyffert, the historian who details early African beekeeping, admitted that very little is known about African beekeeping.[47] In Gambia and Guinea, for instance, Seyffert states, "The absolutely massive abundance of bees, exquisite honey, and equal quality wax allowed a very lively trade to arise very early [in the colonial period]."[48] The Masai tribe sometimes paid as much as two cows for a pot of honey.[49] Angola was noted for having massive quantities of beeswax, so much so that a German ship physician comments on its abundant wax markets as early as 1611 and an Italian writer named Cavazzi reports that beeswax was sold in "masses" as late as 1687. Next to slaves, beeswax was the second-leading commodity in the region.[50]

Yet no records exist that suggest Africans became beekeepers upon their arrival in the New World, although thousands of slaves were brought to the colonies during the seventeenth century, and even more during the eighteenth century. They would have had to adjust to a different kind of honey bee. However, given the agrarian nature of colonial life, slaves would have had to use the products of honey bees, and in later centuries, their ancestors would appropriate the image to make their own comments about American life in their literature and music.

For many Native American tribes, America was never a biblical land of milk and honey. America was nonetheless a complex, interwoven economic continent in which tribes engaged in trade with other tribes in order to supplement their needs. It was a highly formalized and centralized continent, in which "the [tribal] links are apparent in the spread of raw materials and finished goods, of beliefs and ceremonies, and of techniques of food production and for manufacturing."[51] Corn, beans, and squash—often called the three sisters—provided the basic diet. Salt, fish, and venison were other important commodities among the American Indians. Maple syrup was the primary sweetener, so Indians did not need honey.

However, the Eastern Woodland Indians had cultivated forests and farmlands that enabled European plants and insects such as the honey bee to thrive in the new continent. Historian Stephen Potter describes the slash-and-burn techniques that many Algonquian tribes used to control underbrush and weeds: "The Algonquians prepared their fields by girdling the trees near the roots and then scorching the trunks with fire to prevent any further growth."[52] Once dead, many large trees were cut down, but many were also left standing. Thus, the remaining stands were perfect hives for the honey bees that survived the Atlantic crossing.

Few records of Indian reactions to honey bees exist from this time period. As previously mentioned, the Swedes brought black bees (also known as German bees) with them in 1638. These bees are mean-tempered, even by today's standards, and so it is little wonder that the Indians associated the black bees with the bullets used in the settlers' guns.[53]

In a popular story told in Massachusetts, Indians were able to see that Europeans used the honey bee as a symbol of hard work and industry. Bee historian Frank Pellett recorded a tale that goes something like this: "When the Massachusetts colony was first beginning, a little story is told of a local Indian, who curiously observed the bees at work. He had seen the horse, and ox—animals previously unknown to him and his people. He marveled that they should toil at the command of the settler. The honey bee was also a stranger, and the situation seemed to have become serious. 'Huh! White man work, make horse work, make ox work, now make fly work; this Injun go away.'"[54] In this story, the Native American is not characterized as "idle," but neither does he value industry as the English define it. Even though Indians would incorporate honey into their diet and trade, they were not prolific honey producers.

They never accepted the rhetoric of being "busy as a bee," although Thomas Morton was one of the few that compared the Native American methods of food storage to that of the "industrious 'Ant and Bee.'"[55] This tale was an early indication of the major difference in value systems. With the beehive model still being a formative influence in the seventeenth century English social philosophies, the colonists modeled their settlements on the principles of industry and thrift—not wanting to waste time, money, or energy. However, Indians, who were not beekeepers and who had not been subjected to various amounts of literature about honey bees, did not restructure their societies to accommodate the disciplined work schedules or labor divisions of the English. The resulting stereotypes the English developed about Native Americans, then, suggested moral flaws in Native Americans that were false, but that justified poor treatment of the Indian people and their lands.

Although North American plants that provided pollen were widely scattered, colonial honey bees did well. In fact, according to Alfred Crosby, "The immigrant insects did as well or better than the Europeans themselves in seventeenth century British America."[56] The colonists brought German bees, which were used to cold temperatures in Russia, Poland, Scandinavia, Germany, France, and the British Isles. The temperatures in North America were comparable to those regions, and thus the bees did well. There also was a variety of nectar and pollen-producing plants: maple, elm, blackberry, tulip tree, basswood, sumac, tupelo gum, goldenrod, aster.[57]

Colonial honey bees could be found in three types of shelters: straw skeps, bee gums, or bee trees. As a result of Eastern Woodland Indian agricultural practices, the North American forests contained healthy stands. "Many woodlands in North America then contained mature trees with large cavities in which bees could nest, and honey hunting was common," states Crane.[58] Black gum trees were especially prone to leaving strong cavities once the tree died, and thus, the term *bee gum* developed. A bee gum is defined as a colony that lives in a hollow log.

Using straw skeps, the English colonists brought German bees, or the "dark bees," to the New World in 1621. In terms of storing the bee colonies for the voyage across the Atlantic, there are few records. But according to Crane, the voyage would have taken six to eight weeks. Quoting an 1830 book that described how cargo from Antwerp was

1.6. Bee gum, courtesy of Berea College Appalachian Museum. Early bee hives in America were called "gums" because black gum trees tended to decay from the interior quickly. This form of beekeeping was practical for frontierspeople, but was banned by the twentieth century in favor the Langstroth hive, which was considered more sanitary.

arranged, Edward Goodell provides a good description: "[The hives] were placed on deck as follows: A strong oak platform was built on the stern of the ship, the crate containing the skeps was securely bolted to this platform facing the sea at the rear of the ship. This kept the bees as far as possible from the ship's crew, and passengers, so that both could go about their business, neither interfering with the other."[59] The bees were brought in straw skeps, which had several advantages. Bees could stay warm in the winter and cool in the summer. The skeps could also be transferred easily: the early seventeenth-century models had handles on them.

Until sugarcane became an established commodity in the Barbados and the American colonies, honey was an important sweetener, medicinal liquid, and brewing agent in the colonies for the same reasons it was important to England.[60] "Maple sugar and sirup, honey and molasses were the only sweetening agents that might be generally available," explains historian Everett Oertel. "Of these only honey could be used without some form of processing."[61] Furthermore, according to Jordan and Kaups, honey supplemented the frontier diet with a "high-energy food at a sensitive and vulnerable time of the year, the spring."[62] Candles also were a necessity in frontier America. Beeswax was the preferred material because of its odorless and smokeless flame. It was a pleasant alternative to cow tallow, bear tallow, or whale oil.

But Crane asserts that the American beekeeping practices were more primitive than those skills that had been used for thousands of years in the Mediterranean.[63] While very sophisticated hives were being developed in England, colonists were content to keep bees in gum stands or straw skeps.[64] In fact, eighteenth-century historian Ann Withington verifies the unsanitary agrarian practices: "Englishmen who traveled in America expressed their horror at the wastefulness and backwardness of American farmers."[65]

These primitive practices had a price. Eric Nelson speculates that after 1670 beekeeping declined because of American foulbrood.[66] American foulbrood was and remains a contagious bee disease. Spore-forming bacteria cause it, and the bacteria can remain inactive for as long as thirty years in the hive and on equipment. Under the right circumstances, the spore will be become active when bee larvae are three days old or younger. The larvae will eat the spore. The spore then grows like a mold, consuming the larvae. The bees recognize the dead larvae and try to clean out the hive, but in the process of cleaning, the healthy bees spread the bacteria. The only known control until the twentieth century was to burn the hives and bees. Although it had been a worldwide problem for beekeepers, including those arriving from England in the seventeenth century, American scientists figured out the cause of the organism once microscopes were invented. However, during the colonial period, American foulbrood swept up and down the Atlantic coast quickly.

Nonetheless, legal documents verify that honey and beeswax were important commodities in the colonies. Because hard money was in short supply and was often sent back to England to pay for agricultural wares, colonists often used "country pay," in which farm commodities were used to barter for living essentials.[67] Honey fit into the "country pay" category. Villages ensured that honey would be part of the economy. As previously mentioned, Newbury and Danvers had their own apiaries, and even Jamestown enjoyed a cottage industry in beekeeping by 1648.[68]

The value of honey often depended on the coin denominations of the colonies' motherlands. For instance, Delaware colonists traded in terms of Swedish dalers and skillings. The French would trade in livres, sous, and deniers, and the English in pounds, shillings, and pence sterling.[69] Peter Force's "Tracts on Virginia" reported in 1650 that honey was sold at 2 shillings a gallon and beeswax at 4 shillings per hundred pounds.[70] Colonists in Virginia, New York, North Carolina, and Connecticut left beeswax and/or bee gums to their children or family relations in their wills.[71]

Even though American foulbrood was a problem at the end of the seventeenth century, bees continued to swarm in America. Both by natural dispersal and by colonists, the honey bees went west along the waterways, which at the time extended to North Carolina, Georgia, and finally, during the early eighteenth century, Kentucky and western Pennsylvania.[72]

The foremost bee historian Eva Crane once remarked that before the 1800s there was very little to be appreciated in American beekeeping.[73] Colonial beekeeping was remarkable, however, because of how pervasive English ideas about bees, especially drones, were applied to colonists. In both Jamestown and Massachusetts, the biblical ideas associated with honey were incredibly important to helping early settlers deal with depression and homesickness. The biblical promise gave colonists hope that North America would give them what Europe could not: land, a second chance, a church of one's own. Once colonists owned land, they appreciated the economic opportunities not afforded them in England.

Furthermore, the American colonies contained diverse communities. Bees fit in well with Massachusetts governor John Winthrop's vision of a "City upon a Hill," a place where Christian values merged with the Puritan work ethic to create a culture in which industry and efficiency would equal financial reward and respect. The Swedes, the Africans, the Native Americans, the Germans—all complicate the picture of colonial New England being a "new Canaan," however. While some groups were proficient beekeepers, other groups did not define industry, associated with the beehive, as a social virtue. The beehive rhetoric, which was impressed on the white Judeo-Christian communities in Europe and during the first stages of New World settlement, was difficult to comprehend by those people who were not English. Unfair land acquisition, poor treatment, disenfranchisement—that is, the real consequences of the beehive rhetoric—were easier to understand and would have disastrous consequences for those not welcome in the City upon a Hill.

Part Two

Establishing a New Colony

Chapter 2

Bees and the Revolution

> The bees have generally extended themselves into the country, a little in advance of the white settlers. The Indians therefore call them the white man's fly, and consider their approach as indicating the approach of the settlements of the whites.
>
> —*Thomas Jefferson,* Notes on the State of Virginia, *1785*

Eighteenth-century America is noted for three interrelated but complex processes: European immigration, frontier migration, and political independence from England. Just as eighteenth-century American society was an intersection of ethnicities, so too was the honey bee a symbol for intersecting, and at times conflicting, values. European immigration had continued unabated since 1683. Anxious to throw off the yoke of state-sponsored religions, many Protestant groups—Moravians, Quakers, Lutherans, Separatists—continued to arrive from Germany and England. These European immigrants brought beekeeping skills. As they continued westward from Pennsylvania into Kentucky, Ohio, Indiana, and Illinois, they took their skeps and bee gums with them.

Although many immigrants kept bees in straw skeps, many frontiersmen were honey hunters. North America offered vast, healthy forests with few government restraints. Even Native Americans incorporated beeswax and honey into the frontier barter system that existed between slaves, French traders, and English, German, and Dutch settlers. Two writers of diverse backgrounds—Moravian missionary David Zeisberger and Louisiana plantation owner Antoine Le Page du Pratz—wrote journals documenting the existence of bee trees and white-Indian exchanges.

Although these trade patterns were interrupted during the Revolutionary period, the honey bee was an important symbol for moderation in the emerging American society. The Poet of the American Revolution, Philip Morin Freneau, wrote a poem in which a honey bee needs to learn restraint, offering a veiled lesson to the Americans just acquiring freedom from England. Frenchman John Crèvecoeur wrote a pastoral in which America, "the most perfect society," closely resembles an industrious hive. Other documents featuring a hive symbol, such as money or calling cards, were widely distributed throughout the colonies both before and after the American Revolution. "Image makers used the visual media for numerous political purposes," explains Lester Olson, "many of which had little to do with overtly picturing the nation as a body politic but nonetheless contributed to the creation of a body politic by inculcating a revolutionary mentality."[1] The 1779 Philadelphia Continental Congress adopted the bee skep on its currency, and after the war, social clubs used the beehive image to promote order and organization. In general, eighteenth-century America depended on a variety of inextricable and intertwined networks, and from the colonists' perspective, the bee was a benign symbol connecting them.

Although bees had already begun swarming west, pioneers carried hives and social customs involving bees into new territory, primarily along established water routes. Colonel James Harrod was one of the first people to bring honey bees into Kentucky in 1780, but bees had already existed in the Appalachians before his arrival. According to noted beekeeper George W. Demaree, "Kentucky in her early history was famous on account of her wonderful forests. In those days many persons kept bees in tall log gums and boxes, and the bees succeeded in propagating the species and bearing up under the disadvantages imposed on them by their ignorant keepers in a manner which would put the best of the races to blush under like treatment at the present day."[2] Frontier people embraced the bee skep image nonetheless. A cabinet made during this time period in Kentucky features a bee skep in its center.[3]

Named in honor of King George III, a new English colony was established in 1733 by Perceval, Oglethorpe, and Associates. Georgia had a twofold purpose, according to historian William Sachs: "The colony was to serve as an armed buffer zone against the Spanish in Florida and as a place where debtors could be given a fresh start."[4] By the time the Moravians moved to Savannah, they found that bee trees were already there. By 1770, honey bees had spread to Natchez, Mississippi.[5] According to historian Everett Oertel, wild honey bees were already established in Alabama by 1773.[6]

Furthermore, records suggest that bees had been established in Florida at least as early as 1764. The Spanish Governor of Louisiana Antonio de Ulloa kept extensive journals of his tenure from 1766 to 1768. When Spain ceded Florida to England in 1763, many Spanish inhabitants moved to Havana, Cuba. According to Ulloa, these Florida exiles took their hives with them, "which they set up in Guanavacoa and in some ranches as objects of curiosity; the latter multiplied so swiftly that they spread throughout the hills and it was learned that they were beginning to damage the plantations of sugar cane on which they lived."[7] Before arriving in Louisiana, Ulloa had lived in Cuba from 1764 to 1765 and had thus witnessed firsthand the rapid dispersal of bees. In fact, he continued to predict that "each bee hive swarmed once a month, sometimes twice; the one regular size, the other smaller, . . . and in the wax and honey taken out there were no less abundant than is the case here [Spain], where this phenomenon occurs only once or twice a year." Ulloa suggested that honey bees could be "advantageous for the national trade," provided, of course, that bees would not replace sugar cane as the main commodity.[8]

Farther north, Pennsylvania had already adopted the bee as its symbol of thrift and industry. And for good reason. By 1771, 29,261 pounds of beeswax were exported from Philadelphia.[9] Many of the German immigrants were skeppists, and their social customs involved bees as well. When a beekeeper died, the Germans believed that the bees must be told. According to folklorist Lester Breininger, the German phrase is "*Won en eama mon starbed, mus ebber die eama ricka,*" or "When a beekeeper dies, someone must inform the bees."[10] If the bees were not told, they would either leave or die.

The English also brought variations of this custom to the colonies. According to folklorist Wayland Hand, "The announcement of death is made verbally, either in a loud voice, as in New Hampshire, or by whispering, as in Suffolk, Leicester, Rutland, and Kentucky." The verse sung in New Hampshire should be rhymed: "Bees, bees, awake! / Your master is dead, / And another you must take."[11] This tradition, according to twentieth-century writer J. E. Crane in *Bee Culture,* must have had its roots in the simple fact that bee colonies might be neglected after the beekeeper's death.[12] Another social custom associated with a death in the family is a process known as "ricking."[13] The eldest son would move the beehive slightly to the right; this movement would signal to the bees that the universe had changed, ever so slightly. In some regions, people dressed the hives in black mourning fabric. Bees were thought to be averse to profanity and quarreling families. "Quarreling about bees also keeps them from prospering," according to Hand, and "these same beliefs are found in Ontario, upper New York State, and in Maryland."[14]

Many pastors were beekeepers, often supplying their own wax for religious ceremonies or using honey to supplement their incomes or diets. Reverend Noah Atwater provided one of the first journals about his garden in Westfield, Connecticut. He recorded swarming patterns and made metheglin (fermented honey wine).[15]

Metheglin (a type of mead) is a beverage handed down from medieval Druids, but it was just as welcome to farmers in eighteenth-century America. Pennsylvanian historian Breininger provides an updated recipe: "To some new honey, add spring water, three parts water and one part honey. Put an egg into this. Boil the liquor till the egg swims. Strain, pour into a cask. For every fifteen gallons add two ounces of ginger and one of cinnamon, cloves, and mace, all bruised and tied up in a sack. Accelerate the fermentation with yeast. When worked sufficiently, bung up. In six weeks, it should be drawn off into bottles."[16] Mead, honey, and beeswax were all items that Reverend Atwater used in his home and village. Because Atwater routinely gave up part of his salary to help the poor in his town, he may have supplemented his income or diet by the honey and wax taken from the bees.

The English, French, Spanish, and Indians engaged in extensive trading and bartering on the frontier. According to L. R. Stewart, "In 1793, on Black River (lower Wabash), a French trapper, Pierre De Van, came upon a party of Indians and two white captives . . . and all were carrying vessels of different kinds filled with honey."[17] Stewart also documents that Indians used propolis for waterproofing canoes, honey for cooking, and beeswax for barter.

The Moravians recorded the honey and beeswax trade with Indians with fastidious detail. The Moravians, a pacifist group, immigrated to America when they could no longer practice their religion freely in Germany. They refused to take oaths to any political government. They also embraced celibacy and the divine lot system, believing that God was ultimately in charge of all decisions, including whom one should marry. During worship, they used the purest materials available, and thus in 1747, their American ceremonies had to have beeswax. Candles were a mainstay in their worship practices.

As business people, Moravians were ethical. Products were produced and sold at cost, which worked well in the manorial system of Germany and in frontier America. Because Moravian communities distinguished themselves by their pacifist stances and piety, their sponsor, Count Ludwig von Zinzindorf, had no difficulty obtaining charters to establish colonies in Georgia, Pennsylvania, and North Carolina. Mission work was always a high priority among the Moravians. Much of the best missionary work done with Indians was done by the American Moravians in the eighteenth century. In their journals, Moravians recorded bee trees, trade negotiations, and the changing frontier.

In Georgia, for instance, Moravians wanted to educate slaves and Indians near Savannah. En route to visit the Indians, Commissary von Reck and Reverend Bolzius found honey in some bee trees; they used it along with "Parrots and Partridges," which "make us here a very good dish."[18] The missionaries were not content with the results of their attempts at converting blacks or Indians, however. Georgia plantation owners did not want their slaves to be taught to read and write, nor did they want the slaves converted to Christianity. The ever-present threat of a Spanish invasion from Florida was also a problem. So the Moravians abandoned the Georgia settlement and relocated to Bethlehem, Pennsylvania, and Salem, North Carolina.

The Moravian colony in North Carolina was more successful than the one in Georgia. The Moravians quickly established a thriving town with woodworking industries, farms, and pubs. They maintained peace in spite of the animosity between the North Carolina backwoodsmen and British Tories in the region. Historian Hunter James describes the potential for conflict on the North Carolina pre-Revolutionary frontier:

> Most of the men and women who had settled on the Carolina frontier had come south looking for cheap land and a chance to be free of the galling restraints—the boundary disputes, high consumer costs, high taxes, and overweening crown authorities—that had plagued them in the North [Pennsylvania]. The Moravians, however, had come for different reasons: for cheap land, yes, but also to live as a "quiet people," to conduct themselves peaceably at whatever cost, and to spread their gospel throughout the length and breadth of a region that had known the footsteps of few hunters and trappers and fewer missionaries.[19]

By providing supplies such as candles, beeswax, and hives to Tories and settlers alike at reasonable prices, the Moravians managed to strike a balance between the Tories and the settlers. A case in point: British Governor William Tryon. An aristocrat, William Tryon was "the unwitting embodiment of just about everything the [American] backswoodsmen despised in their British overlords."[20] However, Governor Tryon had a very good opinion of the Moravians because they refused to join any of the riots, which began the buildup to the events of the 1770s. On one visit in 1767, Tryon ordered 478 pounds of candles, 150 pounds of butter, 6 beehives, 3 bushels of rye, a gun, and even a windmill.[21]

After the Revolution, Moravians expanded westward into the Ohio frontier, establishing missions in both Gnadenhutten, Ohio, and Schoenbrunn, Illinois. Although the landscape was filled with bee trees aplenty, Ohio was bloody and violent ground. The Moravians were in territory disputed by the Indians, British, French, and American forces. Trying to strike an uneasy balance between these forces was a young Moravian named David Zeisberger, who had been trained in Savannah, Georgia.[22] He had studied the Mohawk languages with Christopher Pyrleaus.[23] Attuned to the subtle changes taking place in American culture and landscape, Zeisberger recorded many Indian-colonist exchanges in the eighteenth century and even passed information to American soldiers about British movements. Of all the men on the Ohio frontier, he was best poised to see the most potential for Native American–Christian communities and best understood the far-reaching implications when that potential backfired.

In 1775, after extensive negotiations, Zeisberger had convinced the Delaware Indians to become Moravian Christians. In 1776, he called for those Christian Indians who were living in Gnadenhutten and Schoenbrunn to move to Lichtenau, the "pasture of light." Although these missions had been thriving and the crops were close to harvest, the Moravians and Delaware Christians left, bringing the cattle. With his missions consolidated in Lichtenau, Zeisberger could see his ideal vision within reach: "A different spirit rules among the Indians. We have seen many who in times past, were our bitter enemies and would neither hear nor know anything of God's Word and now show themselves to be very obliging and confiding toward us."

But the American Revolution broke out, and "unfortunately for the Christian Indians, their villages and missions lay between the hostile British and their allies and the Americans," explains Ohio historian R. Douglas Hurt.[24] In 1778, the non-Delaware Indians strengthened their alliance with British forces. In 1781, the British decided to move the Moravians to the Sandusky, where the commandant at Detroit, Major Arent Schuyler De Peyster, could be sure that the Moravians were not giving information to the American troops. The move to the fort proved disastrous for the Indians. While in transit, Zeisberger wrote, "We felt the power of darkness, as if the air was filled with evil spirits." In 1782, with his people close to starvation, Zeisberger agreed that some of the Delaware Christians could return to Gnadenhutten to harvest the crops they had left in the fields.

More than ninety Delaware Moravians were picking corn in Gnadenhutten on March 8, 1782, when a Pennsylvania militia rode up. When the militia called to the Indians to leave the fields, the Delaware went willingly into the houses. The militia began to slaughter them. The Indians sang hymns throughout the massacre: "The militiamen took the Indians in groups of two or three to the two cabins that served as slaughterhouses, made them kneel, and smashed their skulls with a cooper's mallet, not unlike the way they would kill oxen or hogs."[25]

Zeisberger stayed in Ohio after the Gnadenhutten massacre, but sadness pervaded his outlook from then on. Perhaps concerned about the Delaware Indians, who were always hungry once their lands were signed away, Zeisberger mentions honey and bee trees frequently in his later diaries. In 1786, he mentions that the Cuyahoga River at the mouth of Lake Erie had bee trees: "The bush swarms with bees."[26] By 1792, a Moravian missionary settlement of Delaware Indians was established in Fairfield (east of Detroit, now in Kent County, Ontario).[27] In an entry dated June 27, 1793, Zeisberger mentions two hives of bees brought there from a Moravian mission in Ohio by Indian Peter (Chief Echpalawchund).[28]

Zeisberger, a courageous man placed in difficult circumstances, died with regret. Although he genuinely believed an Edenic paradise could be found in America, he had seen many Indians succumb to European diseases, massacres, starvation, and alcoholism in spite of the Moravians' efforts to help them adapt to Christian religion and customs. His diaries are one of the best resources to study how valiantly a missionary tried to document America as the biblical land of milk and honey, even as its promise failed repeatedly.

Honey bees were in Tennessee as well; they were one of the reasons why Moravians Abraham Steiner and F. C. von Schweinitz found that region so attractive. They reported that pioneers on Judge McNairy's plantation had burned forty swarms of wild bees, which had settled on the garden fence and begun to rob their tame bees in September 1798.[29]

The military also recorded the subtle changes in the landscape and the Native American adaptation to them. Although peach trees were not indigenous to America, the Cherokee tribes readily adopted them. When Colonel Benjamin Hawkins traveled across Tennessee in 1796, he noted the Cherokee had "peach trees and potatoes as well as native corn and beans, and some had bees and honey and did a considerable trade in beeswax."[30]

We learn from existing missionary and military records that Indians incorporated bee trees and honey into their trade and diet patterns very quickly. Perhaps they did so to participate in the barter system. Perhaps they did so because their lands were signed away, and they could no longer count on the traditional "three sisters"—corn, beans, and squash—to sustain them through the difficult winters. We do not have written records documenting precise reasons why, but when describing the complex food and fur trade exchange economy in the Lower Mississippi Valley, historian Daniel H. Unser Jr. explains the importance of the complex barter systems during the eighteenth century: "When one follows the movement of deerskins and foodstuffs through this network, the importance of small-scale trade among diverse groups of people comes into focus."[31] Unser's research legitimizes trade among frontier communities that kept few account ledgers and balance sheets, privileging cultural values more than money or wampum. Even though the Indians may have initially thought of the honey bee as the white man's fly, it had become their fly too.

Ironically, more documents exist about Indians and honey bees than about blacks and honey bees, even though Africa had enjoyed a rich tradition of honey hunting. German historian Carl Seyffert's 1930 book, *Bees and Honey in Africa,* was published after World War I, but he documented colonial beekeeping traditions in African countries, such as Ethiopia, Senegal, Angola, and Gambia. Eva Crane, condensing Seyffert's work as well as translating his sources, begins her chart with an explorer named Alvarez, who described Ethiopia in 1576 as "the whole land . . . overflowing with honey."[32] In 1602, Pierre de Marees recorded honey bees in a tree nest in Guinea. "There is much honey in the forest," Samuel Braun wrote in 1625 about Angola, and the bees "make honey and wax from many sweet plants."[33] Because Africans from all of these countries were shipped to America during the seventeenth and eighteenth centuries, I suggest that they brought their honey hunting skills with them, although I have been unable to find records to prove this link.

West African slaves were especially important to the still-struggling American colonies. In South Carolina, slaves had rice skills that were necessary on plantations.[34] So many slaves from Angola were brought into Louisiana that d'Iberville named his plantation after the West African country. Even in New York, the "Dutch Colonial" house owners depended on slave labor to quarry the stones and cut the timber.[35] The proprietors of New Jersey granted farmers 75 acres for every slave brought into the colony. With so many slaves from West Africa, America was "a nation within a nation" by 1706, according to scholar Margaret Washington.[36]

And yet the only reliable American narrative documenting the importance of honey in Africa is *The Interesting Narrative of the Life of*

Olaudah Equiano (1789). Equiano, also known as Gustavus Vassa, was born in Nigeria in 1745 and sold as a slave to a ship captain in the British Royal Navy.[37] Although there is evidence to suggest that he was really born in South Carolina, Equiano provides an important portrait of pre–slave trade Africa in his narrative: "Our land [Africa] is uncommonly rich and fruitful, and produces all kinds of vegetables in great abundance. We have plenty of Indian corn, and vast quantities of cotton and tobacco. Our pineapple grow without culture; they are about the size of a large sugar-loaf, and finely flavored. We have also spices of different kinds, particularly pepper; and a variety of delicious fruits which I have never seen in Europe; together with gums of various kinds, and honey in abundance."[38] Most probably compiled from oral narratives, Equiano's narrative nonetheless documents that many slaves in America used honey in their African diets and gift-giving rituals.

American slaves often made candles. Although everyday candles were made from a variety of substances (such as tallow, sperm oil, and bayberry), beeswax was the preferred material for candles used during important celebrations. An important rice plantation in South Carolina, Middleton Place (built in 1741), indicates that, in its task system, three slave women made beeswax and bear tallow candles throughout the year. Middleton Place was a showpiece among the colonial elite, and beeswax candles would have been used for its elaborate ceremonies.

Farther north in Virginia, George Washington, another slave owner, had hives on his plantation. Martha Washington reputedly liked rose-flavored honey. The following is a recipe provided by the estate: one should bring a cup of mild-flavored honey to a boil in a heavy saucepan. Turn off the heat as soon as the honey starts to foam up. Stir in half a cup of fresh rose petals. Let the mixture sit for four hours. Bring to a boil again. Pour through a strainer and discard the petals.[39]

Farther west, French Louisiana had plenty of bees and an extensive food network that brought together whites, blacks, Cajuns, Creoles, and Indians. In his journal, Antoine Le Page du Pratz, director of the Chapitoulas sugarcane plantation in Louisiana, allowed his slaves to supplement their diets by harvesting their crops to sell at the local food markets. He was so pleased with his slaves' participation in the food markets that he highly recommended that other directors let their slaves have small plots of land, known as concessions.[40] Slaves were able to sell their extra produce and gain a small degree of autonomy from their owners. Nor were concessions limited to just a few slaves. According to Daniel Unser, "Many of the several thousand African slaves shipped to Louisiana during the 1720s to expand the commercial agriculture turned to small-scale cultivating and marketing of foodstuffs."[41]

When Louisiana became a state in 1812, American authorities cracked down on the slaves' freedom in the markets. But they could not control the daily cross-cultural interactions that occurred between the Spanish, Indian, French, and Africans in the markets and on the El Camino Real (also known as the King's Highway or the San Antonio Trail), a trade route that had been established since the 1500s.[42] Because slave women were responsible for meals, they were a steady presence in the Louisiana food markets, buying and selling surplus items for their owners and themselves. "From these circumstances in the marketplace, not to mention those in the colonial kitchens," Unser states, "came the heavy African influence upon Louisiana's famous Creole cuisine."[43]

Furthermore, the El Camino Real was a major artery for travel and trade between Spanish Texas, the French town of Nacogdoches, and the Caddo villages on the Natchez during the seventeenth century. Begun first by the Apache and Tonkawa Indians in the 1500s, the Spanish used it to establish mission and trade posts in the 1700s. The best description about the complexity of this seemingly simple trade route is found in the Handbook of Texas Online: "Although generally thought of as one road, it may be more accurate to say it was a network of trails—different routes used at different times," and I might add, by different peoples.[44]

Because many Africans could not read or write English, recipes were handed down orally. Census records indicate that slaves made candles, kept gardens, and cooked in the kitchen. These sparse records are a tenuous link between American slaves and their honey traditions in Africa. "Clearly, Indians did not just hunt, blacks did not just grow crops for export, and whites did not merely choose to become either subsistence farmers or staple planters," states Unser.[45]

Antoine Le Page du Pratz was not the only Frenchman observing bees on his Louisiana plantation. The most charming honey hunter in America, in my mind, was a Frenchman living in Nantucket. St. John de Crèvecoeur was fiercely loyal to King George III but passionately in love with America. Written from the viewpoint of Farmer James, *Letters from an American Farmer* (1781) links the honey bee with efficiency, economy, and egalitarianism in America.

Crèvecoeur, trained in manners and sensibility, embraced the customs of rural life with sensitivity and gusto. He preferred the locust bee trees to the "polished mahogany hives" found in Europe. In fact, Crèvecoeur provides one of the best descriptive passages about how to hunt for honey. "I make a small fire on some flat stones, in a convenient place; on the fire I put some wax; close by this fire, on another stone, I drop honey in distinct drops, which I surround with small quantities of vermilion, laid on the stone; and then I retire carefully to watch whether any bees appear."[46]

Crèvecoeur also describes the legalities concerning bee trees: "If we find anywhere in the woods (no matter on whose land) what is called a 'bee-tree,' we must mark it; in the fall of the year when we propose to cut it down, our duty is to inform the proprietor of the land, who is entitled to half the contents; if this is not complied with we are exposed to an action of trespass, as well as he who should go and cut down a bee-tree which he had neither found out nor marked."[47] This pastoral, then, privileges an orderly, law-abiding frontier. In this new classless society (which Crèvecoeur called the "perfect society"), every farmer had a chance to participate in owning land, and there were laws to protect his property.

Crèvecoeur gently parodies the stereotypes of hunters and their game by poking fun at himself and his chase of the "harmless bees": "I cannot boast that this chase is so noble, or so famous among men, but I find it less fatiguing, and full as profitable; and the last consideration is the only one that moves me." This man wants none of the glory of hunting big game. He cares not for boasting rights. In fact, he admits that the first bees he found were by accident! In many ways, Crèvecoeur extends the seventeenth-century English metaphors of the New World as a new hive. Ever the optimist, Crèvecoeur equates New England with a bee colony, calling Nantucket a "fruitful hive constantly [sending] out swarms, as industrious as themselves, yet it always remains full without having any useless drones."[48]

Yet America was not the perfect society, as Crèvecoeur well knew. Crèvecoeur provides one of the most scathing portraits of slavery in Charleston, South Carolina. Furthermore, the Revolution would force him to return to France, never to return to America or his family. But his pastoral was immensely popular in Europe. Thus, in linking Nantucket with bees, he perpetuates a metaphor of America as an ideal beehive while at the same time suggesting that this paradise needed to examine the institution of slavery.

The American Revolution was the formative moment of the eighteenth century, and writers funneled their energies into political treatises, tracts, and poems. More precisely, writers used the drone bee image to change how American colonists perceived British tax collectors. In order to enforce the Stamp Act and the Townsend Acts, England sent tax collectors and administrators to the colonies. According to Ann Withington, "Americans labeled the officials 'drones' who 'did no work of their own, but lived off the work of others' (a not unreasonable assessment given the exorbitant fines that officials pocketed for trivial, technical, often unavoidable violations of the acts)." The officials were drones in part because they did not do anything except attempt to collect taxes levied by King George III.[49] The Stamp and Townsend Acts infuriated the colonists; they wanted no part of another English war with France or to be taxed without representation in Parliament.

When the American colonists decided to revolt, they wanted to forge a new identity, and establishing a new currency was an important mark of independence. The Continental Congress of Philadelphia (1779) adopted the bee skep to put on its currency. Even though Continental Congress bills were considered worthless, the skep was important for the young nation trying to imagine itself as an independent country. Given the high illiteracy rates and unstable state-sponsored banking systems, colonists needed to have an image that would offer stability. Robert Garson explains the importance of the early banknotes: "Money was an artefact as well as a trading device. . . . They were visual reminders of the connection between finance, stability, and national authority. In both function and design money promotes specific political preferences."[50]

Although used only briefly, the bee skep with thirteen rings promoted political stability during uncertain times. The 1779 Continental currency was marked by a red wax seal, which showed that it was an official bill. The wax seal was used to undermine British efforts to counterfeit the Continental currency. A British sloop christened the *Phoenix* was anchored in the Philadelphia harbor throughout the early years of the Revolutionary War. On board was a team of British counterfeiters

2.1. Continental currency from 1779, courtesy of the Smithsonian Institution, NNC, Richard Doty. The Continental Congress endorsed the bee hive as a sign of political stability. The skep has thirteen rings, each symbolizing a colony. A red wax watermark, which was Benjamin Franklin's method of foiling the British counterfeiters, is visible. In fact, the phrase "not worth a continental" developed during this period.

determined to drive up inflation. Even though they succeeded in their efforts to increase inflation, colonists continued to trade, barter, and work together to defeat the British and claim independence.[51]

The bees also served their country during the Revolution. The story goes like this: The British army under the leadership of General Cornwallis had been planning an attack on the ragged Revolutionaries, but no one knew when the attack would take place. One day, while walking down a lane, a young Quaker girl named Charity Crabtree saw a wounded soldier coming her way. The soldier gave Charity this message: Cornwallis's army will attack on Monday. The soldier asked the girl to take the message to General George Washington and then collapsed before the oncoming British soldiers. The girl mounted the horse and began to gallop away, but she quickly realized that her chances of outrunning the Redcoats were slim. Thinking quickly, Charity upset her bee skeps; the bees immediately attacked the soldiers. She galloped off to General Washington, who credited her with saving the country. "It was the cackling geese that saved Rome," he is supposed to have said. "But it is the bees that saved America."[52] It's a fun story for several reasons. It juxtaposes a simple Quaker girl with a sophisticated general, and it shows that women are just as capable of strategy as the military leaders. Most importantly, the bees are given credit for being the protectors of home and hearth.

After the American Revolution, many social clubs emerged to take advantage of freedoms and new advances in the sciences. Manufacturers, merchants, and farmers had cards engraved with the beehive to symbolize industry. According to Ann Withington, "In post-revolutionary America . . . a hive symbolized a society working together for the good of the whole, whatever the nature of the whole might be."[53] These engravings depict women carrying bee skeps and suggest that in the New Eden women would be an integral and affirmative part of the new democratic order. The women stare from the drawing, strong, relaxed, and as comfortable with a skep as they are with a shield.[54] In other engravings, hives imply social order. Often, the hive dominates the center of the page. If the hive is off to the side, it is balanced by another image. In keeping with Enlightenment principles, American engravers maintained order and balance in their artwork, even though the new states were hardly organized into a cohesive country.

The most famous of these societies was the Freemasons. As a secret society, it was a way for men to network and to advance in the fledgling republic. Many prominent men were members, including George Washington, Paul Revere, and John Adams. In many speeches given by the Freemasons, honey bees were used to encourage industry and thrift, a tradition that continued until the twentieth century. In his 1920s farewell address, for example, Grand Master O. D. Street incorporated bees: "Look well, my Brethren, to the traditions of our operative Masons of old, that in their loyal observance of the lessons taught by the Bee Hive, we may find inspiration for a new and continuing devotion to the ideal of work."[55]

Street left his brethren with a wish that someone would discover "a remedy for the 'drone-evil.'" If a brother could find such a cure, "He would place the whole Fraternity under everlasting indebtedness to his genius. The bees kill their drones, but that would be an unhappy manner

2.2. Certificate of membership for the Society for the Promotion of Agriculture, Arts, & Manufacturers (New York). Courtesy of Winterthur Museum. This association was started immediately after the Revolutionary War and reflected a commitment to order, moderation, and knowledge. The hive's central location on the certificate pulls together the various elements of society: shipping, timber, agriculture, sciences, law, and the arts.

of disposing of ours. How to destroy 'dronishness' without killing the drones, as Hamlet would say, 'that is the question.'"[56]

Even though Street was writing in the twentieth century, he was borrowing from eighteenth-century images, values, and sermons held dear by the Freemasons. Arnold and Connie Krochmal show a skep on a Masonic document engraved in 1772 in Newburyport, Massachusetts.[57] Grand Master O. D. Street provides the clearest reason why the emblem was perfect for a new fraternal society: "In all Nature there is nothing more constantly busy than the bee," he says. "It has been the emblem of diligence since antiquity. No symbol of labor could be more appropriate

2.3. Certificate of membership for the Salem Charitable Mechanic Association. Courtesy of Peabody Essex Museum. This association's motto, "Let Prudence Govern, Fear Not," emphasized moderation and control as the country was in transition. Overseeing the developments is Archimedes, generally regarded as one of the greatest mathematicians of all time. The bee skep is featured next to the coin, emphasizing commerce. The shipping and printing industry are also featured in this scene, stressing the balance and confidence of the new nation.

than the bee hive, the abode of great industry. Masonry signifies labor. Toil is noble. Idleness is dishonor."

When the Revolution ended, printing presses were able to print information about beekeeping in America. Until the Revolution had ended, the first books in America were British books that had little familiarity with American soils, plants, or trees. "The texts began their Americanization by being printed in the colonies," states Philip Mason, "not by being written by colonials."[58] In fact, American beekeepers do not profit from local publications until "after the censorship of pre-Revolutionary

War was lifted and the rapid spread of printing presses after the American Revolution allowed for the publication of information on topics of interest to the citizens of a new and agrarian nation."[59]

In his 1998 dissertation, Philip Mason credits George Cooke for writing the first book about bees and beekeeping in America. In his book *The Complete English Farmer: Or Husbandry made perfectly easy in All Its Useful Branches. Containing what Every Farmer Ought to Know and Practice* (1772), Cooke devotes an entire chapter to beekeeping. Cooke's chapter records the English peasant customs that formed part of the early American beekeeping rituals: "It is general custom, when the swarm is risen, to make a noise with a pan, kettle, mortar, and c. but some reckon it an insignificant ceremony, and others esteem it prejudicial."[60]

This old European custom, also known as tanging, actually began by an English king's decree, according to Roger and Kathy Hultgren: "In England, the King declared by law, whenever a swarm of bees emerged, the beekeeper would drum on tin pans or ring a bell. This signaled that the bees that were in flight were his and he was the only one able to claim them."[61] Gradually, as the custom was passed down through the years, it took on a new significance. People believed that the noise caused the queen to become confused and her bees to cluster around her. When English and German beekeepers immigrated to the colonies, they brought tanging with them; this tradition lasted well into the nineteenth century.

George Cooke also advised eighteenth-century readers to wash their hands and face with beer when swarming hives, although the reason is not discussed.[62] This book was later published under another title, *The New Complete English Farmer, or, The Whole Body of Husbandry Made Perfectly Easy. Containing what every farmer ought to know and practise, in All The Various Useful Branches of Husbandry.* The two books are similar in substance and syntax. With these exceptions, according to Mason, many agricultural books were not adapted to the American landscape but were merely extensions of British agriculture.

Mason thus corrects a common misperception that Isaiah Thomas had written the first bee book, *A Complete Guide for the Management of Bees, Throughout the Year.* Although the manuscript was indeed printed in Worcester, Massachusetts, in 1782, it was plagiarized from Daniel Wildman's book published in Britain in 1773. Clearly, by the eighteenth century, people were wanting to educate themselves about honey bees, even if they had to rely on English books.

Americans continued to use the term *bee* to describe joyous occasions for fellowship and work, something they had done since colonial days. The *Boston Gazette* officially recognized its usage on October 16, 1769: "People had gathered for a social function, which people in the Countrie called a bee."[63] However, Irving speculated that it began with the Dutch in New Amsterdam in the late 1600s.

Even "The Poet of the American Revolution," Philip Morin Freneau, wrote about honey bees. Although his poem "To a Honey Bee Who Hath Drunk Too Much Wine and Drowned" (1806) has often been considered a parody of the lofty metaphysical conceits and sad elegies common during the eighteenth century, I think it also serves as a warning to Americans intoxicated with political freedom for the first time. The poem begins with a speaker's address to a honey bee circling his glass of wine. In a series of rhetorical questions, the speaker ponders the reasons behind such a visit:

Thou born to sip the lake or spring,
Or quaff the waters of the stream,
Why hither come on vagrant wing?—
Does Bacchus tempting seem—
Did he, for you, the glass prepare?—
Will I admit you to a share?
Did storms harass or foes perplex,
Did wasps or king-birds bring dismay—
Did wars distress, or labours vex,
Or did you miss your way?—
A better seat you could not take
Than on the margin of this lake.

The speaker considers a range of reasons—from Greek gods to natural calamites to poor direction—that would entice the honey bee to the glass of wine. He finally concludes that the "lake" in the wineglass is as suitable as any found in nature.

The speaker then issues an invitation, and a brief warning to the honey bee, who is not accustomed to the merits of wine:

Welcome!—I hail you to my glass:
All welcome, here, you find;
Here let the cloud of trouble pass,
Here, be all care resigned.—
This fluid never fails to please,
And drown the griefs of men or bees.

The cautionary tone urging the honey bee to show some restraint with this unfamiliar liquid is subtle but true. The speaker gives examples of people (that is, "bigger bees") whose love of wine caused them misery.

Yet take not oh! too deep a drink,
And in the ocean die;
Here bigger bees than you might sink,
Even bees full six feet high.
Like Pharaoh, then, you would be said
To perish in a sea of red.

Finally, when it is clear that the honey bee intends to drink too much wine, the author renounces any responsibility for the honey bee's actions, and even provides a proper farewell by sending him to the other world.

Do as you please, your will is mine;
Enjoy it without fear—
And your grave will be this glass of wine,
Your epitaph—a tear—
Go, take your seat in Charon's boat,
We'll tell the hive, you died afloat.

The speaker carefully paces the actions so that the reader imagines the honey bee sipping too much wine, until it finally falls into the glass, too inebriated to save itself. Although written during the nineteenth century, Freneau's poem belongs in this discussion of the Enlightenment because it emphasizes moderation. The irony is that while his poetry reflects conventional themes of balance, Freneau's political writings suggest radical changes in government and political thinking. The marvel of his poem, then, is that Freneau uses a honey bee to show moderation, when normally bees show industry and perhaps even an excessive devotion to work.

Thus, during the eighteenth century, the honey bee was not restricted to one region, one community, or even one value. As the colonies banded together to form a united social front, the beehive image symbolized a stable society in many engravings. Just as interesting, too, is who was not in the engravings. Blacks and Native Americans established trade and food routes as the colonists migrated west, but these ethnic groups were not included in social organizations. Still, Indians and blacks were honey hunters, cooks, candle makers, farmers, and gardeners who participated in the complex frontier barter system, even as larger social forces were marginalizing them.

Finally, American writers used the honey bee in markedly different ways than did their literary predecessors. The literature of the time did not promote adherence to a monarch or a divine being. In fact, Crèvecoeur, Jefferson, Zeisberger, and Freneau suggested literary independence from England. (These writers were not European.) In their writings, honey bees were stripped of their ancient associations with Olympian gods or Christian saints. Enlightenment principles of moderation and industry were exemplified in the honey bee, although the drone image still carried negative connotations. These writers provided an important foundation, then, for the bee-hunter stereotype in nineteenth-century literature.

Part Three

Swarming West during the Nineteenth Century

Chapter 3

Before Bee Space 1801–1860

> The Indians with surprise found the mouldering trees of their forests suddenly teeming with ambrosial sweets, and nothing, I am told, can exceed the greedy relish with which they banquet for the first time upon this unbought luxury of the wilderness.
>
> —*Washington Irving,* A Tour on the Prairies

American beekeeping history is generally divided into two periods: before and after Lorenzo Langstroth. Before Langstroth little was known about how the bee colony functioned. American beekeepers were at the mercy of two phenomena: a disease known as foulbrood and the bees' natural instinct to swarm. Once Langstroth invented a hive that was compatible with how bees built wax combs, however, beekeepers could take better care of and profit from their hives. Because his discovery happened in 1851, a chronological division has been convenient for historians to use as a demarcation point when discussing bee history. The interracial and international social networks that existed between immigrants, pioneers, and Indians before Langstroth's research has been overlooked because of how quickly the bee industry developed once the concept of bee space was invented.

At the beginning of the nineteenth century, America seemed an unlikely place for the beekeeping innovations that would revolutionize the world by the end of the century. The smoker had yet to be invented, so many beekeepers used sulfur to kill bees before taking honey. Furthermore, American foulbrood destroyed many bee colonies. Although foulbrood was serious (the spores can remain dormant for eighty years), it was not the only threat to bees. The *Boston Patriot* published an account of wax moth for the first time in 1806.

Although there are two kinds of wax moth, greater and lesser, this chapter refers to the greater wax moth, which has been a pest for a longer time. To paraphrase Roger Morse, wax moth larvae destroy honeycombs by boring through the wax in search of food.[1] Strong colonies can withstand wax moth by evicting the larvae, but especially in the South, bees are almost always dealing with this pest. The German black bees, which could survive the colder temperatures in New England, were very susceptible to wax moth. Since a beekeeper could not check colonies for moths when bees were kept in straw skeps or bee gums, very little could be done to prevent the spread of moths. In the words of current *Bee Culture* editor Kim Flottum, "Wax moth is like a bad cold. [Contemporary] Beekeepers will always have to deal with it, but it is not especially disastrous if proper steps are taken beforehand to keep it from getting out of control."[2]

However, during the early nineteenth century, beekeepers had very little control because hives that they could open on a systematic basis had not been developed at this point. In fact, historian Wyatt Mangum suggests that wax moth might have been a problem much earlier as a result of the method used to ship bees across the Atlantic. Because straw skeps were placed in large crates, the chance that moths would be nested in the hives would have been not only possible, but probable.[3] In any event, within two years of the 1806 *Boston Patriot* article, four-fifths of all apiaries in the Boston vicinity were abandoned.[4]

Meanwhile, the beekeeping community in Europe was rapidly unlocking the secrets of the hive. Or so it seemed. A Swiss scientist, Francois Huber, designed an observation hive in spite of being blind. With the help of his wife and another assistant, Huber recorded his insights in his

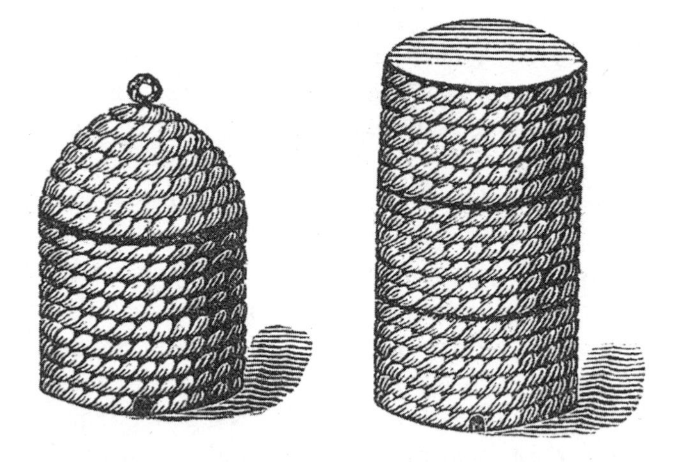

3.1. Storifying using cylindrical skeps and a bell-shaped top skep. Originally illustrated in Bagster (1838). Courtesy of Kritsky, May 2003. Elaborate skep architecture was designed to facilitate the honey bees' natural tendency to cluster in the winter and provide a place for bees to store extra honey. Neither invention was as effective as the Langstroth hive, which was more efficient in terms of honey storage.

Letters, which was distributed among beekeepers everywhere, including America. Furthermore, Johann Dzierzon, a pastor in Silesia (now Poland), was also developing his own theories about beehive construction. Elaborate hives and skeps were constructed. Glass jars placed on the tops of skeps were really the first types of observation hives.[5] Three-story octagonal hives—first designed in the seventeenth century—underwent major innovations in the nineteenth century so that they produced more honey. According to Gene Kritsky, beekeepers wanted to build hives that worked with the bees' natural rhythms. So, using an eighteenth-century beekeeper named John Thorley as his source, Kritsky explains, "The use of the octagonal shape was thought to be the closest to a circle that could be produced from flat pieces of wood or glass, and this was the ideal shape to help the bees survive the winter, as they naturally clustered in a circular mass."[6]

Furthermore, European beekeepers stayed in contact with one another more than beekeepers in America did. They had libraries, shared resources, and enjoyed better postal systems than Americans did before the 1800s. Because many beekeepers were clergymen, they had access to universities, scientists, and archives that American beekeepers could only dream about.

When compared with Europe, American beekeeping was unsophisticated at the beginning of the nineteenth century. In fact, if the bees did survive American foulbrood and wax moth, American beekeepers would kill the weakest colonies at the end of a season, knowing the bees couldn't survive the winters. The strong colonies were also destroyed because these would yield larger quantities of honey. Only the midweight colonies would be saved for the following year. This practice alone was enough to make American beekeepers seem cruel and primitive by European standards.

By the end of the nineteenth century, however, America would become the leading influence on beekeeping throughout the world. "What had been needed all along," explains Florence Naile, Lorenzo Langstroth's biographer, "was a hive so built as to let bees live and work in their own way, to give their keeper ready access to the interior for inspection and care, and to permit the removal of surplus combs without waste of honey, disturbance of the colony, or injury of worker-bees."[7]

Once Langstroth developed a hive with moveable frames, beekeeping progressed in the following ways: beekeeping changed from a cottage industry to a profitable industry; bees benefited from better care; and large-scale commercial agriculture of tree crops and vegetables transformed America's landscape and role as a political power. Furthermore, in a culture that was beginning to develop its own leisure habits, honey hunting remained a profitable sport in America.[8] The same powerful media forces that educated emerging beekeepers also perpetuated the stereotype of the bee hunter and circulated instances of Old World folk practices such as tanging in the new country. Newspapers and books were much more common than in the previous centuries. These media forms perpetuated the sport of honey hunting, and European journals promoted beekeeping. Both activities emphasized powerful but contradictory values: independence and environmental sustainability on the one hand; scientific progress and efficiency on the other.

The contradictory values associated with honey bees paralleled the values associated with westward migration. In 1803, Thomas Jefferson sealed the details on the Louisiana Purchase, opening up land that would eventually become fifteen states.[9] The Northwest Ordinance brought peace to the Ohio Territory and established Jefferson's policy regarding treaties with Native Americans—that is, "bribery, not war." Jefferson's policy was effective at maneuvering the Indians further west. Only one year later, Chief Keokuk was persuaded to sign over an additional 15 million acres, which included states from Missouri to Wisconsin.

From this point on, the prairie lands were an open invitation to German, French, and Irish immigrants. These immigrants differed from those who participated in the initial colonization of the seventeenth century. Many were disillusioned with the revolutionary efforts in the European countries, displaced by industrialization, and frustrated with European land controls and high rents. In other words, immigration had less to do with religious freedom than it had for immigrants of thc colonial period. These nineteenth-century immigrants primarily sought economic opportunities, but they were still challenged by frontier conditions. So many economic, cultural, and physical benefits of honey and beeswax were still appreciated by these settlers, quite a number of whom came believing in the travel literature that promoted America as a paradise.

Compared with the civil unrest gripping Europe, America *was* a paradise for people, slaves being the notable exception. But bees suffered. Although wax moth had plagued beekeepers since the time of Aristotle and Virgil, it was not native to the United States. Once wax moth appeared during the nineteenth century, the disease spread as quickly as American foulbrood had during the colonial period. Early records from this time provide a clue that overcrowding of hives could cause bee diseases. In Trumbull, Ohio, Jared P. Kirtland noted the intense competition between German beekeepers. By 1828, however, wax moth had ravaged their apiaries, and in 1831, every region in Ohio had been affected.[10] Competition between beekeepers was a relatively new phenomenon because most people did not own enough land or resources to care for many hives in Europe. But when swarms were available, American farmers exploited bees as a natural resource as much as they did the land. Furthermore, Crane explains, "The large numbers of honey bee nests in trees contributed to the moth's rapid spread."[11] Beekeepers were desperate. Within the time that the wax moth appeared until Langstroth invented his hive, at least six hundred patents for "new" hives were filed, many claiming resistance to wax moth.

People relied on honey hunting, especially as the Appalachians, Midwest, and West developed. Sequestered from accessible trade routes, the Appalachian people developed a particular fondness for sourwood honey. The Native American tribes incorporated honey into their trading and celebrations. The English botanist Thomas Nuttall observed that the Shawnee Indians near Memphis, Tennessee, had venison and honey on New Year's Day in 1819.[12]

But bees had swarmed by natural dispersal further west than Tennessee. Explorers Meriwether Lewis and William Clark found bees at the Kansas River in 1804, and English naturalist John Bradbury claimed to have seen them in Omaha in 1811. In one trade, Colonel J. W. Johnson, owner of a trading post in Burlington, Iowa, "purchased beeswax and tallow to the value of $141 from the Indians" in March 1809.[13] Records suggest that after Colonel Harrod carried bees into Kentucky in the 1700s, the bees swarmed naturally and prolifically in the Midwest and Plains during the early 1800s. Still, the Plains were as formidable an environment to cross for honey bees as they were for white settlers. Although it is likely that bees would have moved west by their own dispersal, settlers streamlined the process by carrying bees with them in covered wagon into the western territories.

Quite a few men became professional bee hunters during this time, and their exploits—recorded by James Fenimore Cooper, Washington Irving, Silas Turnbo, and Alphonso Wetmore—emphasize the importance of time and money. In these early nineteenth-century narratives, bee hunters were notoriously independent and had followed no one's time schedules but their own—until the hunt began. Then a bee hunter's reputation and career depended on how quickly he could find a bee tree. In these narratives, a good bee hunter could find a bee tree in thirty minutes or less. James Fenimore Cooper perpetuated the bee hunter type in both America and Europe with his wildly popular *Leatherstocking Tales,* especially *The Prairie,* which was published in 1827. The fictional characters were based on true explorers. According to historian Roger Welsch, "These [bee] men were often the most careful explorers of the new frontier country."[14]

Washington Irving's Oklahoma bee hunter, always on the outskirts of civilization, always male, was a fashion faux pas: "A tall, lank fellow in homespun garb that hung loosely about his limbs, and a straw hat shaped not unlike a bee hive; a comrade, equally uncouth in garb, and without a hat, straddled along at his heels, with a long rifle on his shoulders."[15] Frontier beekeepers lived off the land, and thus the rifle was the new element to the typical portrait of the beekeeper. Although he was perfectly capable of working with wild bees, the beekeeper needed more protection on the Oklahoma frontier.

In the lesser-known region of Arkansas, Silas Turnbo recorded at least seven stories about bee hunting in *Chronicles of White River,* several focusing on a local legend named Bill Clark in the Ozark region. Bee hunter Clark had a fine reputation for settling clients' honey and wax requests and kept all of his promises, but he did it on his schedule. Sometimes several months would pass before the honey showed up on a doorstep. Turnbo also included stories of unethical bee hunters, giving the stereotype an edge that had not appeared in the genre before.[16] In Turnbo's chronicles, unscrupulous bee hunters—such as those who would steal or chop down their neighbor's bee trees—were part of the reality of frontier ethics. But Turnbo also included a method describing how hunters would drink honey out of a deer leg: "The usual way of making the case was to take the hair off the hide and thoroughly tan it and sew it all up into a sack except the end of one leg which was left open. Sometimes though the hide was formed into a case without removing the hair. When the case contained honey and a hunter become hungry for a drink of honey, he would suck it from the aperture in the leg."[17] A successful hunt was judged by how many deerskins were filled with honey. Turnbo had a fastidious eye for details, and his chronicles provide valuable information to contemporary audiences wanting to know the nitty-gritty details of an old-fashioned honey-hunting party.

While Turnbo wrote mainly about Arkansas, Washington Irving wrote about Oklahoma as he traveled with Charles Joseph Latrobe's 1832 expedition. Compared with Turnbo, Irving idealized "The Honey Camp" and "The Bee Hunt" in *A Tour on the Prairies* (1835), glamorizing the Great Plains as a place where one could live off the land.

The honey camp is described as a "wild bandit or Robin Hood scene . . . in a beautiful open forest, traversed by a running stream, where booths of bark and branches, and tents of blankets." The Creole cook on the tour suggests a diverse community existing on the Plains: "Beatty brought in a couple of wild turkeys; the spits were laden, and the camp-kettle crammed with meat; and, to crown our luxuries, a basin filled with great flakes of delicious honey, the spoils of a plundered bee-tree, was given us by one of the rangers."[18] Not content to merely describe the bee camp and its cook, Irving also describes a bee hunt. Noting how Indians have reacted to the abundance of bee trees, Irving states: "It is surprising in what countless swarms the bees have overspread the Far West within but a moderate number of years. The Indians [the Osage and Pawnee were in Oklahoma] consider them the harbinger of the white man, as the buffalo is of the red man; and say that, in proportion as the bee advances, the Indian and the buffalo retire."

Irving provided one of the fullest discussions about how Native Americans included honey into their diet and trading patterns: "The Indians with surprise found the mouldering trees of their forests suddenly teeming with ambrosial sweets, and nothing, I am told, can exceed the greedy relish with which they banquet for the first time upon this unbought luxury of the wilderness."[19] The Indians were already well-established buyers and traders when discussing weapons and food; Irving's compliment acknowledged the Indians' ready acceptance into their "banquet."

As if to accentuate the juxtaposition between the orderly bee colony and the Oklahoma frontier, Irving describes the bees as urban professionals, unaware of the impending downturn in their forest marketplace: "The jarring blows of the axe seemed to have no effect in alarming or disturbing this most industrious community. They continued to ply at their usual occupations, some arriving full freighted into port, others sallying forth on new expeditions, like so many merchantmen in a money-making metropolis, little suspicious of impending bankruptcy and downfall."[20] Irving's parallel between the businesslike bee community and an emerging market class was one of the finer additions since bee literature started with Virgil's *Georgics*. Irving added the element of money, linking it to honey, and thus producing a very capitalistic—and very American—spin on the bee swarm and the bee hunter.

No less of a frontiersman than Davy Crockett was pleased to find that in Texas "there were bees and honey a-plenty" in 1836. A bee hunter named Gideon Lincecum wandered through the Texas Republic in 1835:

> I lived plentifully all the while. . . . Every time I found honey I would have a feast of the first order. I could kill venison any time, and to broil the back-straps of a deer on the coals, dip the point of the done meat into the honey, and then seize it in your teeth and saw it off with your knife is the best and most pleasant way to eat it. I have often thought that there could be no other preparation of food for man that is so suitable, so agreeable, and so exactly suited to his constitutional requirements.[21]

Of these honey hunters, the legendary Tom Owen stood out, as much for what he didn't do as what he could do. He did not lure bees to a flour-covered plate and then follow them the way many of his peers did. This process, which is still used in some parts today, is called "coursing" a bee, and because the bee is covered with flour, a hunter can easily follow it back to the colony. Instead, Tom Owen depended entirely on eyesight to find trees over a mile away. Owen would say, "These clever little beasts have their set ways. One has only to watch the air current. When they go out foraging they always fly against the wind, then the breeze helps them to carry home their heavy load."[22]

From Texas to Iowa, bee hunters on the plains were a marked contrast to the calm, quiet monks or subservient peasants who were generally associated with European beekeeping, subjected to rigid laws and taxation, and confined to unbending social classes and economic opportunities. Too, since printing presses had become mechanized, the bee hunter became a type that was marketed just as the Indian was: the man pushing the boundaries of the frontier, one foot in the wild, one foot in civilization, independent, isolated, and loving only the order of the hive. In early nineteenth-century narratives, the bee hunter moved on the margins of a society, searching for a liquid gold rather than cities of gold. In the increasingly secular American West, that was as close as the bee hunter would come to a church.

Even as these bee hunters were living off the land, settlers were changing the land in ways that would end bee hunting as a professional career. Plows meant that settlers would till up the prairie wildflowers, and trees were cut down to make houses. As the open range was enclosed for more refined cattle than the Texas longhorns, the pieces were in place for a beekeeping industry to develop later in the century. Bees and cattle had followed one another since Greek and Roman antiquity, and when the European settlers began invading North America, this formula did not change, although the unique nature of the American landscape did affect how bees and cattle would be kept. In the nineteenth century, settlers were still figuring out the vagaries of the Great Plains.

In Southwest Texas, the cattle cradle had already formed when the Spanish left their cattle to run wild after their failed attempts to conquer and civilize Texas in their usual four-step fashion: conquest, mission, presidio, and pueblo.[23] The Spanish, always greedy for gold, had not counted on fierce Plains Indians, and eventually, they abandoned their efforts at the end of the eighteenth century. While the Texas region underwent a series of political changes in the nineteenth century—a province, a republic, a state, and a confederate state—the Spanish longhorns reproduced and multiplied rapidly near San Antonio. This "cattle cradle" had water, shade, and forage materials. By the time Texas became an American territory in the 1840s, immigrant beekeepers from Germany were able to capitalize on the region's abundant pollen and nectar supplies. The region seemed to fulfill biblical promises.

Above the Plains, bee hunters and settlers mingled on the Missouri frontier. For many, a leisurely honey hunt was just the diversion needed from the hardships of frontier life. Of these, Dr. Josiah Gregg was the most famous. His *Commerce of the Prairies* advocated that honey hunts were good for one's health.[24] Although honey hunting could be beneficial to humans, more than a few honey hunters were unethical when it came to taking care of the frontier forests. They could be "a people less industrious than the insects which they destroy," to quote nineteenth-century journalist Alphonso Wetmore.[25] He claimed that Clinton County, Missouri, was filled with "church-going citizens—except when snake-killing and bee-hunting were in order."[26] In Harrison County, Missouri, whole neighborhoods would participate in honey- and wax-gathering expeditions.

Conscientious pioneers were angry about the wastefulness of trees as early as 1821. Henry R. Schoolcraft, nineteenth-century Ojibwa scholar and friend to Henry Wadsworth Longfellow, wrote a particularly "scathing account of the manner in which two pioneers gobbled up handfuls of honey, sharing it the while with their hounds."[27] So, too, did Irving write about the "wasteful prodigality of hunters," explaining that since the "surrounding country, in fact, abounded with game . . . the camp was overstocked with provisions, and no less than twenty bee trees had been cut down in the vicinity."[28] So many people participated in honey hunting that in 1837, Alphonso Wetmore "was concerned about the damage done to forests by bee hunters and tanners." Daniel McKinley reports that eleven bee trees were required to make a barrel, and in one instance in July 1842, a party was able to take seven barrels of honey.[29]

Reports of the plentiful bee trees in the Midwest abounded, however. In 1836, one man in Scotland County, Missouri, reported seeing seventy-five barrels of honey on its way to the market. George Combs found over 170 bee trees in his first season in Clark County.[30] One group of settlers from Iowa found sixteen white oak trunks with bees in 1839.[31] Beekeeping laws, which protected forests and honey hunters alike in Europe, were nonexistent on the frontier. Circumstances were right, then, for an old-fashioned feud to develop between Missouri and Iowa involving the felling of honey trees on the border.

Dulce Et Decorum Est

In 1839, the snow fell deep and blurred the unclear boundary between Iowa and Missouri.[32] The ambiguity about the boundary had arisen from two different surveyor charts, specifically about where the rapids of the river Des Moines started. When the first surveyor went to chart the land, the river was at a normal level, and he used a series of the rapids where the Des Moines met the Mississippi to chart a boundary line. However, a second surveyor went when the river was higher, and when he wrote his report, the rapids appeared to start further north on the Des Moines River than where the first surveyor claimed the usual series of rapids were.

The area was rich in wild game and bee trees, and when Missouri ordered another survey done, the surveyor decided to use the shallower rapids of the Des Moines river, instead of those that appeared where the Des Moines met the Mississippi. Doing so accomplished two things for Missouri: nice straight geographical lines, and more land.

Iowans were not pleased.

As timing would have it, two seemingly unrelated events happen at the same time. When a Missourian cut down three bee trees in the snowy disputed land territory, an Iowa court naturally ruled against him. In an unrelated incident, a Missouri sheriff named Uriah S. Sandy Gregory tried to collect taxes from Iowa people who were building houses in the disputed territory. He was chased out of town in short order.

But these two incidents were enough to get Missouri Governor Lilburn Boggs's blood boiling. Boggs knew a thing or two about chasing people out of town: he later forced the Mormons out of Missouri, at gunpoint. He wouldn't be outdone by a bunch of Iowans. Governor Boggs demanded that Sheriff Gregory go back to Iowa and collect taxes. When Gregory finally did so, Iowa Governor Robert Lucas declared an old-fashioned border war. That was fine by Boggs. To read Sue Hubbell's description of the account, the soldiers in both armies were hardly the noble type: when ammunitions stores were closed, they broke in to steal weapons. When asked for volunteers, no one would step across the line, except for two inebriated men. It was December, after all. Both armies preferred that Congress decide where the border was.[33] A truce was called, a decision was handed down, and when the final tab was calculated, both states had to pay more than they felt that they should have had to. The bee wrangler who started the feud, it seems, went unpunished.

Small wonder, then, that even though Indian tribes adapted to the benefits of the hive, some still regretted the bees' presence. When talking about the Osage Indians in the Missouri territory in 1836, Alphonso Wetmore "claimed that he was once at the Osage village near Papinville when the Indians held a day of mourning because a swarm of bees had been found."[34] It was a sure sign Indians would rapidly lose their way of life.

The Great Awakening

Civilizing a wilderness required great time and energy. When America finally stopped fighting the British, dealing (and in some cases double-dealing) with the French, and "bribing" the Indians, the country experienced a religious awakening, the effects of which still linger. The Great Awakening prompted a spurt of utopian communities, approximately 450 in all. Thomas More's *Utopia,* written in England during the 1500s, promoted a community built on the principles of justice and peace. Ideally, the community would be classless, and everyone would live in relative comfort. Poverty would be eliminated if everyone worked. Nineteenth-century America provided the perfect circumstances for leaders willing to create a utopia, for America provided land, democracy, and unlimited natural resources.

Because it is impossible for this book to provide a comprehensive discussion of all these societies, I will focus on the more prominent communities that directly or indirectly affected American beekeeping. In addition to the United Society of Believers in Christ's Second Appearance (henceforth known as the Shakers), other religious utopian communities such as Harmonists, Icarians, and Mormons (now called the Church of Jesus Christ Latter-day Saints) depended on the honey bee either as a pragmatic foundation to an agrarian lifestyle or a religious icon.

Although the Shakers had been in America since 1774, their communities struggled until the nineteenth century. After Shaker Elder Joseph Meacham wrote official guidelines in 1787, the Shakers agreed to communal living, equality between the sexes, and celibacy in order to focus on spiritual matters. Above all, instead of a Holy Trinity, which many Protestant communities embraced, the Shakers believed in a divine duality that embraced both sexes—the male aspect of the Father and the female aspect of Wisdom.[35] The Shakers believed that God was not found in an afterlife, but in the present activities of daily life.

The Shakers' radical beliefs regarding women and divinity meant that the Shakers were constantly harassed. They were beaten, jailed, and often run out of town. When the frontier started expanding west, the Shakers jumped at the opportunity to create communities in relative isolation from Protestant cities. Still, the Shakers were not unpatriotic in spite of their pacifist philosophies. One Shaker, Issachar Bates, credited George Washington with liberating the colonies from England because such autonomy afforded greater chances at religious toleration in the new country.[36] By 1810, major Shaker communities had been established in Kentucky, Ohio, and Indiana, and all kept bees. In addition to the newer communities, the older, more established communities such as Canterbury, Mount Lebanon, Harvard, Sabbathday Lake, and Hancock kept bees in New York, Massachusetts, and New Hampshire.

Although all Shaker communities depended on agriculture, the Kentucky villages of Pleasant Hill and South Union took great pride in their apple orchards and were willing to experiment with imported bees, bee houses, and beehives. Bee houses were included in the community plans. Scholar Julia Neal writes of Pleasant Hill: "Around each family dwelling house were clustered numerous small dependencies, such as the butter house, the fruit-drying house, the chicken house, the woodhouse, the round bee house, and the silk house."[37]

Their efforts were not in vain. In 1834, two prominent Shaker leaders in the East—Isaac Newton Youngs and Rufus Bishop—decided to tour the villages in the West to observe the new experiments. They were treated like celebrities. In a travel journal written by Youngs, he writes of visiting John Smith's bee house in South Union, Kentucky: "This was a curious site [*sic*] it had a round brick building two stories high and so constructed that one can go into the inside and see the bees to work in their places or boxes."[38] Interestingly enough, there were two John Smiths at the South Union village: one a respected leader, one an "absconder," who drifted in and out for about twenty years before the Shakers lost patience with him. We don't know which man was the beekeeper, but the journal entry shows that the Shakers borrowed vocabulary from the bee world (for example, absconding) to describe the various people who drifted into their communities.

In a bee journal written by Giles B. Avery in 1851 to 1854, he recorded the daily trials and tribulations of working with bees and bee swarms in Lebanon, New York. The cold weather was certainly a factor, as was the amount of light in his bee house, causing bees to want to fly out when it was too cold or causing the hives to mold. Avery's journal is a meticulous account written by a man gathering swarms in orchards, feeding his bees, and buying swarms from his neighbors (both Shakers and non-Shakers).[39]

The image of a drone was used frequently in Shaker hymns. The Shakers borrowed Isaac Watts's English 1655 hymn: "How doth the little busy bee, / Improve each shining hour" to encourage their fellow brethren to renounce earthly desires:

Behold in spring see everything
Alive and cloth'd with beauty
Shall I alone an idle drone
Be slothful in my duty?
To gather honey see the Bee Fly

Around from flower to flower
A good example there for me
To well improve each hour.[40]

Because the Shakers believed that Christ had already come in the form of their leader, Mother Anne Lee, they refused to sing hymns celebrating an afterlife. Such messages, they felt, were misguided.

In contrast to the Shakers, the Harmonians believed that Christ's second coming was imminent. George Rapp, a dissident Lutheran, immigrated from Württemberg, Germany, to establish Harmony, Indiana, which emphasized communal prayer, music, and work. Rapp wanted to create communities that would be perfect models of Christian life, so that when Christ came back, Christ would be more pleased with the Harmonian societies.

As with all frontier communities, honey bees served an important function in this community. Extensive bee skeps were located near the apple orchards, and wax was available in its stores for nearby residents.[41] As a symbol of biblical importance, the honey bee served as a symbol of harmony, even though the community quickly fell into discord when Rapp fell in love with a very beautiful and very young woman.

Amid controversy, George Rapp sold the Harmony community to the wealthy philanthropist Robert Owen, who promptly renamed it New Harmony. Owen had made millions in the English mill industry, but he wanted to start a socialist community. Knowledge, not religion, was his priority. Owen enticed the finest minds to come to America. Although the community ultimately failed because it was so disorganized, New Harmony made lasting contributions to American education and science. When entomology was just emerging in this country, a New Harmony resident Charles Alexander Leseur illustrated a bee tree to include in Thomas Say's *American Entomology, or Descriptions of the Insects of North America*.[42]

Of all the utopian communities in the nineteenth century, the Church of Latter-day Saints best illustrated the successful assimilation of religious beliefs and American ideals embodied in the honey bee. The prophet Joseph Smith, born in Palmyra, New York, received a vision from Moroni that extended Christianity to include American Indians. Having found gold plates containing hieroglyphics in his dream, he translated them into the Book of Mormon. Central to the Mormon religion is the honey

3.2. Charles Alexandre Lesueur, title page for Thomas Say's *American Entomology, or Descriptions of the Insects of North America* (1824). Courtesy of New Harmony Working Man's Institute. This nineteenth-century illustration shows a bee tree sheltering a variety of American insects. The symbols of civilization are in the background; a steamer is in the distance. But in the foreground is a web of insects, starting with the bees, moving into spiders, drifting into butterflies and ants.

bee, also called *deseret* in the Mormon language. Although other Masonic symbols, such as the all-seeing eye, the compass, and the sun, were important to Mormon pilgrims, none would permeate Mormon communities like the bee skep and the honey bee.

When the Mormons followed Smith to Nauvoo, Illinois, in 1838, they embarked on an exodus that kept them emigrating until they finally reached Utah in 1847. Through it all, the beehive was used in iconic fashion to help solidify Mormon identity. Smith's bodyguards, called the Lifeguard Legion, had the skep emblazoned on their uniform buttons. Although the skep certainly embodied industry, it also meant collective safety to the Mormons.

And for good reason. The Mormons were harassed just as the Shakers had been. In Illinois, an anti-Mormon mob lynched Smith and his brother while they were awaiting trial in a Carthage jail in 1844. Upon Governor Thomas Ford's insistence, the Mormons fled Illinois bound for Utah, which was still a territory. They took bees with them. In fact, contemporary commercial beekeepers Russell and Norman Mitchell can claim a great-grandfather "who brought bees to Utah, strapped to the back of a covered wagon, with Brigham Young."[43] The Mitchell family descendants did not stay in Utah, preferring Idaho instead, but they did remain beekeepers.

So did the Mormons. The bee skep became the image that alternately provided protection and encouragement for Mormons. In a popular song "The Busy Bees of Deseret," the Mormons recorded the intense hatred that drove them to the Utah territory in 1848: "The busy bees of Deseret / Are still around their hives. / Though honey hunters in the world / Don't wish these bees to thrive."[44] These bees did thrive, however, in part because of the extensive missionary system set up by Brigham Young when Utah was still a territory. He sent missionaries to recruit artists and skilled craftsmen from throughout the world. A good number of these new recruits were Danes, English, Welsh, and German.

Governor Ford had been anxious to see the Mormons go, and he welcomed the French Icarians, led by Etienne Cabet, who were considered some of the finest minds in Europe, being an eclectic mix of engineers, philosophers, and writers. Committed to a more perfect society, Cabet wanted Napolean to establish a French republic. Exiled for his beliefs when Louis Phillipe of Orleans was crowned king, Cabet went to England, where he read Owen's tracts for New Harmony. Greatly influenced by Owen's writings, Cabet wrote about a perfect communist society in a novel called *The Voyage to Icaria* (so named for Icarus, who flew too close to the sun). Cabet convinced French civilians deluded with Napoleon Bonaparte III's efforts in France to follow him to America.

The first group of Icarians, sixty-seven men, landed first in Texas and made their way to present-day Dallas. According to M. G. Dadant, "The type of men in the party hardly warranted stopping in such a place where no cultivation had as yet been attempted. They had but one farmer in the lot of 67."[45] Predictably, many of their company soon perished, and they relocated to New Orleans. From there, Cabet appealed to Governor Ford (who had just threatened to exterminate any remaining Mormon in Nauvoo, Illinois) for a safer, more civilized place to relocate his group.

Caught in a combination of unrealistic ideals, bad business planning, and a poor work ethic, the Icarians could not survive the harsh Illinois conditions, even though Nauvoo was, compared with other settlements, quite established. Most of the Icarians had given the majority of their funds to Cabet, who was not a good businessman. By the time Cabet had purchased Nauvoo for the Icarians, he only had funds to buy beans for the group, many of whom had come from upper-class backgrounds.

Furthermore, the German Americans in the region could not stand the Icarians. To quote German immigrant Viktor Bracht, who wrote about the Icarians when they lived in Texas: "Communism and Mormonism are diseased plants that will not thrive even in the healthful West nor anywhere else in the New World. There the human spirit tolerates no restrictions. . . . Communism in America, therefore, need not set its expectations too high."[46] Initially, Bracht appeared to be right. The Mormons *had* been shuffled off to Utah, where circumstances for survival, much less long-term successful civilization, did not seem promising. As soon as Cabet had to go to France to settle legal matters, the Icarians split between those who felt it was beneath their station in life to work and those who were expected to use their skills. Eventually, the Icarians got into a massive food fight. The cooks became angry at the elitist attitudes shown to them by some of the Icarian members and rebelled by throwing food, pots, pans, and plates.

Among the Icarians who moved to Illinois was a man named Marinelli, who had been a tailor in France. When he married, he and his wife had a daughter named Marie, but after the disaster at Nauvoo, they relocated to St. Louis with some of their fellow Icarian friends. Others went to California, and some went to Iowa. A few Icarians stayed in Illinois, however, and made a lasting impact on American beekeeping in the following way. In 1857 a former Icarian, Emile Baxter, planted the first vineyard in Hamilton. When a beekeeper named Charles Dadant decided to relocate to America, he chose to relocate to that region because of its strong French connections and his initial desire to plant his own vineyard. Although he failed in this attempt to be a vintner, he soon followed his passion of beekeeping. His son Camille Pierre married Marinelli's daughter Marie later in the nineteenth century. Thus, when Governor Ford banished the Mormons, he initiated a course of events that would profoundly improve American beekeeping in ways he little suspected.

Whereas Mormons simply wanted to be away from the United States to set up Deseret, Texas wanted to be included. The Mexican War broke out in 1846, and Texas finally became a state in 1848. The most famous beekeeper serving in the Mexican War was a Kentuckian named

Gen. D. L. Adair, who led the Fourth Kentucky Regiment in Mexico. He returned to Kentucky to become one of the most respected beekeepers, lawyers, and writers in the region. In addition to helping set up a public school system, he served as the agriculture editor for the *Louisville Ledger* and issued *Progressive Bee Culture* during the 1870s.

When Texas became a state in 1848, shortly before the Civil War, the West truly opened up for many people. For, according to Walter Prescott Webb, the West was more about *how* cattle were kept rather than *if* cattle were kept; this central difference separates the eastern farms from the western. If one would succeed in the cattle cradle, one needed to have fodder and water for the livestock. Furthermore, if an ambitious man named James Gadsden had had his way, Texas would have had a railroad running to California before the Civil War. In league with his friend Secretary of War Jefferson Davis, Gadsden wanted to make the West dependent on the South for transportation needs. When President Franklin Pierce gave Gadsden permission to buy land from Mexico, he made few stipulations. Gadsden had dreams of a southern industrial-agricultural empire stretching from South Carolina to California, linked by railroad.[47]

With the signing of the Guadalupe-Hidalgo treaty in 1848, Mexican general Santa Anna signed over much of present-day New Mexico, Arizona, California, Colorado, Utah, and Nevada. But Gadsden, president of the South Carolina Railroad Company, wanted more. He wanted a last section of southwestern New Mexico and southern Arizona. After agreeing to pay $10,000,000 for a parcel of land as big as one state in 1852, Americans expressed outrage, although the Gadsden Purchase was a part of the much-larger Manifest Destiny movement.

Although it was a beautiful dream, Gadsden did not count on two things when it came to his southern railroad company: the Plains Indians, who could be a formidable force, and the Civil War. In fact, the Indians would rule the region until after the Civil War, during which time the technology was developed to conquer the Plains Indians who did not want to be "bribed."

In 1849, Germans, discontented with their failed revolution and industrialization, began to leave Germany en masse. Its strict laws regarding property ownership also pushed young men to America because only the oldest son could inherit land. German immigrant Viktor Bracht wrote a treatise "Concerning Emigration," describing the poverty in Germany: "One must be totally blind not to notice the alarming increase in the number of poverty stricken. The dire poverty of the proletariat is ranged in extreme discontent against the wealth of the small number of well-to-do. For the present, the property class is enabled to control the property-less."[48] When Texans wanted to populate the new republic, German Prince Solms Braunfels set up a company to send immigrants over.[49] He chose the crème de la crème of German society, "a class of bold, intelligent and enterprising people," as Viktor Bracht wrote in his travel literature.[50]

Among these new scientists, engineers, and teachers was an apiculturist, Wilhem Bruckish from Prussia. He settled in New Braunfels and established progressive beekeeping. A friend of German bee master Johann Dzierzon, the Congregationalist pastor who learned that queens determined the sex of the bees, Bruckish maintained contact with Dzierzon after he arrived in Texas, although records of Bruckish corresponding with American beekeepers do not exist.[51] Probably the unreliable postal system and language difficulties were factors because New Braunfels was a frontier village until after the Civil War. However, in the lists of commodities that Bracht recorded, honey was plentiful, in either bee trees or beehives. In New Braunfels, San Antonio, and Fredericksburg, one could buy five to eight pounds of honey for a dollar. Beeswax cost a dollar for eight to twelve pounds.

Another young man named Christian Diehnelt landed in New York from Saxony, Germany, in 1852, but eventually made his way to Honey Acres, Wisconsin, to establish a beekeeping business. He recognized the innovative technologies as they were occurring, and he was one of the first buyers of industrial beekeeping equipment once it was offered for sale.[52] But Germany was not the only country sending immigrants to America. The Irish also migrated to America beginning in 1849 because of the potato blight that occurred in 1846. Although Irish beekeepers enjoyed a long and storied tradition that rivaled that of Germany and England, many of the immigrants who arrived in America were penniless and sick. When the famine ships were rejected in New York, which had strict quarantine laws, the Irish who recovered were admitted to other ports such as Philadelphia, Pennsylvania; New Orleans, Louisiana; and Savannah, Georgia.[53] The Irish could not buy land immediately and often settled in the poorer sections of their adopted cities and took low-paying positions.

For those Irish immigrants who went to Texas, the War for Independence decimated their communities. They had to move immediately to Victoria or Petersburg for their safety. Compared with the German immigrants, the Irish were more unsettled in Texas for a longer time. However, in *The Historical Story of Bee County, Texas,* historian Ezell Camp emphasizes that the Irish recorded frontier barter exchanges for beeswax near Galveston.[54]

Midwest settlers found gold in the limestone before they found oil in the panhandle. Bee Cave, Bee Creek, and Honey Grove are only a few oft-repeated place-names.[55] But as idyllic as the names were, the living conditions were otherwise. When Ole Rynning created his travelogue for Norwegians, he emphasized honey bees in Illinois, which would appeal to the Norwegian mind-set. In his persuasive *True Account of America,* Rynning wrote of "the prairies, the wild game, the bees, fruit, rivers."[56]

"It is the Prairie," ranted English writer William Cobbett, "that pretty French word, which means green grass bespandled with daisies and cowslips! Oh God! What delusion!"[57] And sure enough, Ole Rynning's account was pure fiction. He had been sold a malarial swamp in Illinois. His entire colony either died or fled. Nevertheless, in spite of being known as "one whose mouth never knew deceit," Rynning wrote convincingly enough to ensure that the Norwegian people would know the beekeeping skills they had in the Old Country would transfer to the New World as well.

Propaganda is an ugly word to use, but early nineteenth-century writers used bees and honey to create an impression that America could right the wrongs of European culture—disease, famine, hunger, overpopulation, unfair land distribution, and political unrest. The technique worked, but democracy and its rewards were not quite as simple as writers suggested. For most immigrants, the American government indeed had provided opportunities and freedoms that European immigrants could only have dreamed about. But even independent bee hunters depended on larger circumstances to provide them with the liberty to hunt bee trees: available land, lax enforcement of laws, easy immigration systems.[58] Those circumstances hardly mattered to the reading public, fascinated with the frontier and the bee hunter's role in it.

Meanwhile, the shy beekeeper Reverend Lorenzo Langstroth quietly

3.3. Lorenzo Langstroth. Courtesy of *Bee Culture*. This Congregationalist minister discovered the principle of bee space, the working area that bees need in which to build their combs to store honey and brood. He constructed a hive in which moveable frames were spaced equidistant from each other; these moveable frames made possible a beekeeping industry. Langstroth also imported Italian bees, the variety most common in America today.

prepared to revolutionize the beekeeping world by examining the mysteries of the beehive in his backyard. Langstroth began beekeeping in 1838, but he found only one book about beekeeping, Edward Bevan's *The Honey Bee, Its Natural History, Physiology, and Management* (1838), although he was very familiar with Virgil's *Georgics*. Langstroth stated that neither publication was appropriate to the conditions in America.[59]

Born on Christmas Day, 1810, in Pennsylvania, Lorenzo Langstroth was indeed a gift to the beekeepers who followed him. He gave the world a concept known as *bee space*, which is the space between frames that bees need to have to build their combs in a systematic order, although he did not use that term.[60] In any modern hive, one can look into a box filled with removable frames. If a space is larger than 3/8 of an inch, bees will fill the space with comb, which is difficult to crack once it has solidified in place. So if beekeepers keep frames spaced equidistantly, less than 3/8 of an inch, they can remove frames filled with honey and replace them with empty frames that bees can build on. Although a fairly simple concept, Langstroth's realization about bee space took the guesswork out of beekeeping. Beekeepers no longer had to keep bees in skeps or box hives. They no longer had to wonder if their bees were suffering from disease or being attacked by moths. They no longer had to kill their bees if they wanted to remove honey; they could simply replace moveable frames of honey.

But Langstroth had studied to be a pastor, not a scientist or apiculturist. Even though he went to Yale College, according to Naile, "there was little if anything in the prescribed course of study which Langstroth took at Yale College that could have prepared him for such a pursuit as that of his later years—the patient observation and study of the life of the honey bee."[61] In school, he roomed with Professor Denison Olmsted, who completed the first survey of geographical and mineralogical resources of North Carolina. But Yale tested Langstroth's convictions, which would serve him well when he had to defend his contributions to beekeeping in later years. When students organized a strike against the school cafeteria for poor food quality, Langstroth alone walked through it. As a pastor and teacher, Langstroth had a phenomenal amount of tolerance and patience.

Further, Langstroth's respect for women defined him as a progressive. Whenever his "head trouble" would begin to disrupt his pastoral duties, he would return to teach school to women.[62] His belief that women should be educated was as radical as his ancestors' ideas about abolition.[63] He also had great love and respect for his wife, with their marriage being a happy one. When he finally moved back to Philadelphia to open a school for women, he also began to keep bees in a log hive in 1838.

It was hardly the time to begin beekeeping. Wax moth had spread quickly to Ohio and the Midwest. In spite of many efforts to find a suitable hive, no one had been able to improve the basic skep. As if the wax moth weren't enough, the two books Langstroth turned to for advice about beekeeping were problematic: Bevan seemed to doubt the existence of a queen bee, and Virgil's *Georgics* was hardly an authoritative manual in American beekeeping. Even the best beekeepers in the region could not teach Langstroth, because "none of them knew enough to drive bees out of their hives, nor used smoke to facilitate their operations."[64]

Yet Langstroth felt an instinctive love for bees. He wrote of the passion he first felt upon seeing a globe, one of the radical experiments then being done with hives: "In the summer of 1838, the sight of a large glass globe, on the parlor table of a friend, filled with beautiful honey in the

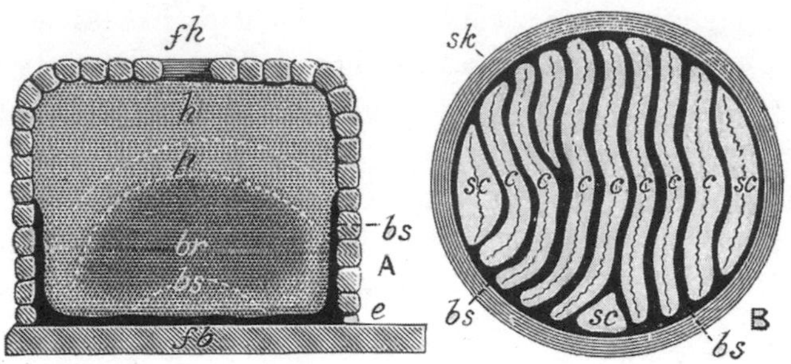

FIG. 7.—STRAW SKEP IN SECTION, SHOWING ARRANGEMENT OF COMBS (Scale, $\frac{1}{12}$).

A, Vertical Section—*fb*, Floor Board; *e*, Entrance; *br*, Brood; *p*, Pollen; *h*, Honey; *fh*, Feeding Hole; *bs*, *bs*, Bee-space. B, Horizontal Section—*sk*, Skep-Side; *c*, *c*, Combs; *sc*, *sc*, Store Combs; *bs*, *bs*, Bee-space.

3.4. The bee space inside a straw skep. Originally illustrated in Cheshire, 1888. Courtesy of Kritsky, October 2003. This illustration provides a look at the concept of bee space (i.e., the 3/8 inch that bees need to have between combs of honey). Not only could beekeepers increase the profits from hives, they could begin regular maintenance of the colonies, which had been impossible with bee skeps.

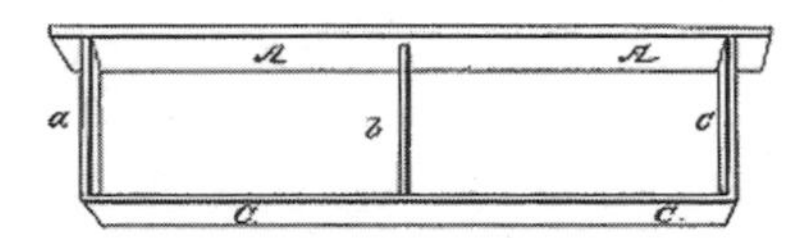

3.5. Langstroth moveable frame. Originally illustrated in Langstroth, 1852. Courtesy of Kritsky, October 2003. When the frame is full of honey, it can be replaced with a new one. Moveable frames made a beekeeping industry efficient and worked with the bees' need for distance between the combs.

3.6. Munn's hive, showing moveable frames for examination. Originally illustrated in Cook, 1884. Courtesy of Kritsky, October 2003. Many people quickly manufactured their own versions using Langstroth's "bee space." Langstroth never received any payment from his patented version of the hive; however, the beekeeping community generously provided funds and aid to him throughout his life.

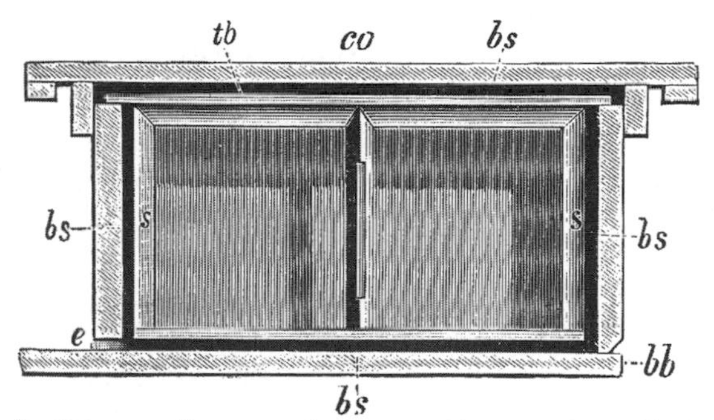

FIG. 11. SECTION OF LANGSTROTH'S ORIGINAL HIVE AND FRAME (Scale, $\frac{1}{10}$).
co, Cover; *bb*, Bottom Board; and *e*, Entrance of Hive; *bs*, *bs*, Bee-space; *tb*, Top Bar; and *s*, *s*, Sides of Frames.

3.7. Langstroth frame showing "bee space." Courtesy of Gene Kritsky. Langstroth never used the term "bee space," but once his discoveries were published, he became the Father of Modern Beekeeping.

comb, led me to visit his bees, kept in an attic chamber; and in a moment, the enthusiasm of my boyish days seemed, like a pent-up fire, to burst out into full flame. Before I went home I bought two stocks of bees in common box hives, and thus my apiarian career began." Fortunately, Langstroth read Francois Huber's letters. One of the great ironies in beekeeping history is that Francois Huber, who was to do so much in the way of an observation hive, was blind.

Born in Geneva, Switzerland, in 1750 and blind from the time he was fifteen, Huber nonetheless advanced beekeeping with the help of his wife, an assistant, and his son. In order to understand how the hive worked, this group devised a leaf hive, which would separate the frames much like the pages of a book standing on end. Huber himself best explains the advantages: "Opening the different divisions in succession, we daily inspected both surfaces of every comb; there was not a single cell where I could not see distinctly whatever passed at all times, nor a single bee, I may almost say, with which we were not particularly impressed."[65] Although Huber advanced the world of beekeeping, he could not offer a practical hive for the average beekeeper.

After his initial experience with box hives, Langstroth obtained two of Bevan's bar hives. In these hives, "the bees constructed combs that hung down from a top bar and were connected to the sides of the hive," writes Gene Kritsky. "The difficulty in working this hive helped Langstroth realize that to have a truly useful hive, he must prevent the bees from attaching the comb to the hive's side."[66]

1850 to 1860: Eureka!

Enough said. Langstroth uttered this word in 1851 when he finally realized that bees need an exact amount of space between the frames. His breakthrough moment occurred on October 31: "The almost self-evident idea of using the same bee space as in the shallow chambers came into my mind, and in a moment the suspended movable frames, kept at a suitable distance from each other and the case containing them, came into being. Seeing by intuition, as it were, the end from the beginning, I could scarcely refrain from shouting out my 'Eureka' in the open streets." This simple discovery revolutionized beekeeping, not only in America but throughout the world. In his patent, Langstroth did not mention bee space by name; he only provided the dimensions: "There should be about three eighths of an inch space between a and c and the sides and C C and the bottom board of the hive [*sic*] this will prevent the bees from attaching the frame to the sides or bottom board of the hive, hindering its easy removal, and will allow them to pass freely between the sides and the bottom board, and the frame so as to afford no lurking place for moths or worms."[67]

Langstroth's "understated" (to use Kritsky's word) description was the most critical aspect of the public's failure to grasp the significance of his discovery. He never reaped the financial windfall that generally accompanies such inventions in America. While he waded through the normal patenting process in 1852, many entrepreneurs had hastily crafted their own versions of his hive and marketed them as their own. When Langstroth tried to reclaim his product through lawsuits, he found that litigation required too much energy. His mental problems would make him an invalid for several months at a time, just what speculators needed to make money off his invention.

The American Dream was just then beginning to be defined by wealth at the expense of exploitable natural resources. Fortunately for Langstroth, his belief in divinity, his sincere love for bees and his fellow neighbors, and his ingenuity brought him enduring fame and respect during his lifetime. Even though financial wealth eluded him, a large community consisting of family, in-laws, church members, and beekeepers cared for and about him. Furthermore, Langstroth continued to revolutionize American beekeeping. He, along with a beekeeper named Samuel Wagner, decided to import the Italian bee, overcoming incredible odds to do so.

In many ways, the Italian honey bee was a perfect fit for nineteenth-century American beekeepers: it could withstand cold temperatures; it tended to be gentler than the common German black bees; and the queens tended to be more prolific.

The logistics of importing the bee were quite difficult given the vagaries of ocean commerce. Beekeepers on both sides of the Atlantic could not figure out how to package the bees so they could withstand the length of the ocean voyage, and sea captains often could not guarantee adequate care in the transatlantic passage. Nevertheless, Langstroth and Wagner finally succeeded in September 1859. The following entries from

Langstroth's diary show how tenacious Langstroth was in his efforts to help the one Italian queen that survived:

> Sept. 4th: Cut out the combs and found a living queen. I never handled anything in my life with such care. If the queen had been killed, I should have felt worse than any regicide ever felt, for they mean to kill royalty. Strange that so small a creature should be capable of producing so exciting an effect!
>
> Sept. 5th: I arose very early, fearful that the queen might have been chilled, and found that the bees had left her. I took her out of the cage with fear and trembling. She was stiff and could hardly move. I warmed her with my breath and returned her to her colony, which I kept in the cellar.
>
> Sept. 7th: Transferred a colony of common bees and removed queen—gave them Italian [queen] in a cage. Near sunset, let out Italian queen very carefully—bees welcomed her. Now I have done all that I could for her and must patiently await the result.[68]

It was only a matter of time until the Italian honey bee became America's darling. After the Civil War, Italian bees were shipped to Texas, Ohio, and California. In 1879, the Italian bee was then imported from California to New Zealand. Other bee varieties excited interest, but according to Eva Crane, "None had such a permanent or important effect on world beekeeping as the Italian."[69]

Of nearly equal importance was a Quaker beekeeper named Moses Quinby. Generous, honest, and a staunch abolitionist, Quinby collaborated with Langstroth on many projects. Of particular importance was his explanation of how to handle American foulbrood in his book *The Mysteries of Beekeeping Explained* (1853). American foulbrood, a bacterial disease, was and is so persistent that it can remain infectious for many years. According to Sue Hubbell, "Bee larvae become infected by eating food contaminated with the spores of the bacteria."[70] Quinby realized that beekeepers needed to start with a hive of fresh frames. No other method to rid foulbrood would be found until the 1940s.

His success as a commercial beekeeper was an important example to

American cottage-industry beekeepers. Honey was his sole source of support from 1853 until he died in 1875. According to historian Philip Mason, "History played its part in the growth of commercial beekeeping: the Civil War shut off the supply of sugar from the southern states, and created a high demand for honey as a sweetener in the North."[71] He died financially secure. In addition to wisdom and books, Quinby gave beekeepers a lasting gift. Shortly before his death, he improved the bellows smoker, making it easier to hold the smoker and control the amount of smoke a beekeeper needs when checking a hive. Similar to the Langstroth hive in importance, the basic principles of the Quinby smoker are still in use by contemporary beekeepers.

Along with Quinby, Adam Grimm best symbolized the American rags-to-riches dream in the beekeeping world. In his history, Frank Pellett makes clear that Grimm did not significantly advance the industry, but he met the challenges of being "a pioneer in a new land with new problems and was making a beginning with a new industry."[72] Born in Germany, Grimm immigrated to Wisconsin in 1849 and made his start with wild swarms that he kept in both boxes and straw hives. Having profited from the Civil War honey prices, he switched to Langstroth hives as soon as possible. When a chance to improve his stock with Italian honey bees arrived, Grimm took it. Ten years later, he sold his crops for $10,000 and started a local bank.

Deseret

Take a stroll through Salt Lake City. The bee skep image is impressed on sidewalk squares, neon signs, doorknobs, and transoms. No other group in the United States has so successfully appropriated the bee skep as its icon as the Church of Latter-day Saints. Although Joseph Smith adopted the bee skep icon in Missouri, Brigham Young branded the image into the Mormon consciousness. The two leaders had one crucial difference, explains scholar Hal Cannon: Joseph Smith wanted to lead a rapid conversion to the new religion of Mormonism; Brigham Young wanted the conversion to be more orderly.[73] For both leaders, however, the bee skep was perfect for the Mormon people: "It was simple, unostentatious, and fitted like a glove," declared none other than Mark Twain in *Roughing It.*[74]

When the Mormons first arrived in the Utah region, they had named

3.8. Mormon currency. Courtesy of the Smithsonian Institution, NNC, Richard Doty. Joseph Smith and Brigham Young used the bee hive to suggest stability and industry. *Deseret* is the Mormon word for honey bee. Although Mormon beekeepers quickly changed to Langstroth hives, the Utah society originally promoted the skep symbol to suggest stability and religious strength.

the place *Deseret,* and according to Carol Simon, the symbol was appropriate: "As a protected structure with no windows to the outside world, the beehive had special appeal to the close-knit pioneers."[75] The lyrics of one song, "All Are Talking of Utah," emphasize the collective nature of Mormonism:

> I now will tell you something you never thought of yet
> We bees are nearly filling the "Hive of Deseret"
> If hurt we'll sting together, and gather all we get
> For all are talking of Utah.[76]

But the traditional values of frugality also appealed to the secular nature of the state. In fact, one writer in the *Deseret News,* in 1881, describes the skep as a "communal coat of arms," which is a "significant representation of the industry, harmony, order, and frugality of the people, and the sweet results of their toil, union and intelligent cooperation."[77]

And toil the Mormons did. Whereas other groups merely crossed over the Rockies to get to California, the Mormons settled in Utah in 1847. "The whole intention of those [wagon] trains was to get an early start, as soon as the grass greened up, and then get *through* the West as fast as possible," explains Wallace Stegner. "The Mormons were an exception, a special breed headed for sanctuary in the heart of the desert, a people with a uniquely cohesive social order and a theocratic discipline that made them better able to survive."[78]

Bee colonies arrived in Utah in 1848. The great-grandfather of beekeepers Russell and Norman Mitchell joined the Mormon migration in Illinois. He strapped the bees on the back of a covered wagon, so the story goes.[79] In addition, none other than Brigham Young himself built the Beehive House, the center of Utah politics, in 1854. A talented carpenter, he carved skeps into the staircase pillars and frames. Bees shine from metal doorknobs, and skeps appear in the wooden doorjambs. In short, every aspect of the house was to be seen as a place of frugality, work, and faith.

However, in order for Utah to be granted statehood, the Mormons agreed to soften the religious implications of the honey bee because Congress wanted a strong division between church and state. The Mormons' petition for statehood was denied six times, just to suggest how much the Mormons had to secularize their petitions. The effect, according to Hal Cannon, was subtle but noticeable: "As [Inspector-General Joseph] Johnston's Army entered Salt Lake Valley in 1858 and the dream of an isolated kingdom ended forever, the symbols and rituals of the church began to take on a new meaning. . . . Benevolent symbols such as the all-seeing eye looked too exclusive to outsiders and were resented by them. . . . Such emblems ran against the grain of the American self-image: independent and free. . . . As statehood came closer, that foreign-sounding word 'Deseret' was dropped from all governmental use."[80]

Still, the skep survived—even thrived. I think there are two reasons for its strength as a symbol. One, the beekeeping traditions in Europe were so entrenched that immigrants arriving from England and Germany latched onto the skep, feeling at home among Mormons at a time when language or cultural skills could pose a barrier. According to a speech "The Mormons" by Thomas L. Kane on discussing immigration to Utah from Great Britain: "They will repay their welcome; for every person gained to the hive of their 'honey state' counts as added wealth. So far, the Mormons write in congratulations that they have not among them 'a single loafer, rich or poor, idle gentleman or lazy vagabond.' They are no Communists; but their experience has taught them the gain of joint stock to capital and combination to labor."[81]

The skep is also less threatening than other symbols. Its rounded corners and its soft colors melding to a conical top suggest maternal inclusiveness. Ironically, the Mormons only used skeps during the early years, but they kept abreast of the changes in technology that affected beekeeping. Mormons used the symbol of the straw skep to promote unity, but they used newer Langstroth hives to earn a profit from honey.

In Walter P. Webb's seminal study, *The Great Plains* (1931), he argues that a lot of industrial inventions aided peoples efforts to conquer the plains: the Colt revolver, a horseman's weapon, to subdue the horse Indians; barbed wire to control cattle; windmills to stock tanks and for irrigation; and railroads to open otherwise unlivable spaces and bring first buffalo hides and buffalo bones, and then cattle and wheat to market, as well as gang machinery to plow, plant, and harvest big fields.[82] Webb did not mention beehives. But Langstroth's model is part of this trend, and the Mormons capitalized on this invention as assuredly as everyone else.

California's Gold Rush

Samuel Harbison, hailing from a beekeeping family in Pennsylvania, first went to California dreaming of gold but quickly changed his mind when he saw there was no future in mining. Harbison could see a future in beekeeping, however, and he had the necessary skills and knowledge to make it happen. Harbison was already familiar with the Langstroth hive, but when he decided to bring his bees west, he modified Langstroth's hive yet again, so that the bees could endure the harsh climate changes that the trip from Pennsylvania, to Nicaragua, to California would entail.

In *History of American Beekeeping*, Frank Pellet credited Harbison with the ingenuity necessary to help bees not only survive the difficult journey, but arrive strong enough to flourish in the new environment once they were allowed to fly again. The crucial difference was minimizing the size of the colony so that the bees had more room than usual. Lee Watkins explains what set Harbison apart from the other emigrants transporting bees across the Rockies: "While the others did little more than buy bees in box hives, tack a wire screen over the bottoms and load them on a boat, Harbison carefully planned his whole operation and devised methods for keeping down the freight costs as well as a way to give the bees plenty of ventilation and extra room for them to cluster in during the tropical heat."[83] On November 30, 1857, the *Sonora* landed with sixty-seven colonies of bees in San Francisco. Harbison made a handsome profit from his hives, selling all but six for $100 gold apiece. He then arranged another trip, adapting the hives again for his purpose: "It was taller and narrower than the Langstroth hive, with a door in the rear opening like a cupboard."[84]

Harbison initiated a beekeeping gold rush on two fronts: smart packaging of bees and smart packaging of honey. Others who tried to imitate his success lost heavily. In fact, historian Stephen Van Wormer discusses an entire region, Pamo Valley, in San Diego County that failed because the environmental factors could not support agriculture. Among those listed as former residents was Ira King, an apiarist, who sold out and went to Texas.[85] But using skill, knowledge, and ingenuity, Harbison overcame the challenges of weather and distance when transporting bees.

North by Northwest

After Harbison's success, it was only a matter of time before honey bees were taken into the Northwest Territory, although it was assumed that the rainfall in the Northwest would not be good for bees. Historian Catherine Williams quotes Charles Stevens, writing in 1853: "There is one thing that I have always wanted to mention, but it has always slipped my mind, and if you ever come to Oregon you must not make any calculations on keeping Bees, for they cannot be raised here, the winters are not cold enough to keep them in, they come out of the hive to fly about, and a little shower of rain will catch them and in that way the whole swarm will be distoryed [*sic*]."[86] The first attempts to bring honey bees to the Northwest were done via covered wagon. According to Williams, Lt. Howison noted in his report than an emigrant of 1846 tried to carry two hives via the Applegate route, but winter overtook them. John Davenport first succeeded in 1854.[87]

As the bees slowly made their way to Oregon, Thomas Eyre emerged as a leader. Reporting that William Buck was selling hives from California, Eyre estimated that the state of Oregon had about 338 hives in 1858. As beekeepers learned more about the weather conditions in Oregon, they began to do simple things to help the hives survive, such as covering them with carpet or matting to protect them from the wind or tipping the hives forward for drainage.[88] When Italian honey bees were introduced, Oregon beekeepers did much better, for Italian honey bees were more resistant to disease than the common German bees.

Spreading the White Man's Foot in Blossom

Henry Wadsworth Longfellow was the first writer to record the Native American reaction to honey bees in literature and thus provide a twist to the American bee hunter stereotype that had been perpetuated by white writers. Longfellow used Iroquois sources to write *The Song of Hiawatha* (1853–1855), but he claimed the book was based on Ojibwa (or Chippewa) myths. The Ojibwa nation had recognized early in the eighteenth century that the honey bee was a symbol of European arrival.[89] In this section titled "White Man's Foot," Longfellow's Hiawatha learns of the white man's advance from the honey bee:

> [Hiawatha:]
> Gitche Manito, the Mighty,
> The Great Spirit, the Creator,
> Sends them [the white men] hither on his errand,
> Sends them to us with his message.
> Wheresoe'er they move, before them
> Swarms the stinging fly, the Ahmo,
> Swarms the bee, the honey-maker,
> Wheresoe'er they tread, beneath them
> Springs a flower unknown among us,
> Springs the White-man's Foot in blossom.[90]

Interestingly enough, in Longfellow's retelling of the Chippewa myth, Hiawatha defeats Megissogwon, the main opponent of the Chippewa, with the help of a trickster figure. But he cannot defeat the advance of white people, nor prevent the destruction of his culture by European ways.

In addition to taking incredible liberties with Indian myths, Longfellow was influenced by the national epic of Finland, the *Kalevala*.[91] Yet even though *Hiawatha* was far from an authentic retelling of Chippewa traditions, the poem was wildly successful both in America and Europe. *The Song of Hiawatha* marks the first time a non-Indian person attempted to set a work within a Native American cultural context. The massive publication of the poem was important because, to quote Joseph Bruchac, it "brought our national consciousness one small step towards the appreciation and acceptance of Native American cultures and Native American literature."[92]

Another development affecting the future of American beekeeping concerns the treaty of the Traverse des Sioux, signed June 1851, which allowed the Americans (also called Long Knives by the Sioux Indians) who had been contained in the Illinois region to claim the Northern Plains states. The treaty allowed Americans to populate 19 million acres in Minnesota, 3 million in Iowa, and 1,750 million in South Dakota. The United States promptly betrayed this treaty; the Indians were never paid for their land, and they later retaliated by killing 800 whites. But as far as American beekeeping is concerned, this treaty opened up land to European immigrants, many of whom became successful pioneers of beekeeping, especially in the twentieth century.

Women also emerged as bee hunters. In the Oregon Territory in 1859, Tabitha Brown organized a school for orphans and depended on honey from her bee tree. "She told of serving her orphans and school children honey from a bee tree. In later years, those who had lived with her verified this," according to Catherine Williams.[93] Brown is the antithesis of the bee hunter: she and her school represented civilization and education. Because education generally follows civilization, American children read nursery rhyme books depicting bees, hives, and skeps during this time period. One of the earliest children's books printed in Boston is *The Alphabet Class* (1848).[94] No illustrations exist to teach the "B" verse, but the content mirrors standard themes associated with honey bees:

B was a Bee that looked well to his hive,
And only by industry labored to thrive.
A wholesome example for you and to me
May even be found in the neat busy bee.

More illustrated resources were forthcoming from Britain. A long poem, "Pretty Name Alphabet," featured in *The Child's Picture Story Book* (1856), depicts a boy named James:

J is for James, to all mischief alive
Who was stung by the bees
When he lifted the hive.[95]

Another British nursery rhyme was popularized in "An Alphabet for Animals" written "by a Lady."

B is the Bee,
So busy and gay;
He is seeking for honey
Sweet honey, all day.
From him to be idle
We learn to avoid—
How happy he is,
For he's always employ'd.[96]

The British press determined how bees would be marketed to children in America during the nineteenth century. Values such as industry and good attitude—long associated with the honey bee—were marketed to American children. The illustrations depicted bees gathering food from flowers, or bees ready to sting. In short, nursery rhymes extended a tradition of using bees to moralize lessons for humans, except this time the audience was children.

During the early nineteenth century, Langstroth's achievements were undoubtedly the most important affecting beekeepers. But other advances tend to be forgotten because so much attention is placed on the progress made possible with the Langstroth hive. The country's settlers made significant inroads—geographically and culturally—using honey-hunting skills to survive in the frontier and to form social relationships with the people they met there. The complexities of the bee-hunter stereotype were finally challenged in literature and on the frontier as more women accompanied men to Utah, Oregon, and California. Unfortunately, these inroads stalled when the Civil War erupted.

Chapter 4

After Bee Space 1860–1900

To make a prairie it takes a clover and one bee.

—*Emily Dickinson*

Compared to honey, sugar always has been a more political commodity, but especially in nineteenth-century America. Until that time, Americans had relied on the sugarcane industry (which had been profitable because of slaves) to serve its collective sweet tooth and had neglected to address the conflict between democratic principles and chattel slavery. The American slave trade had been inextricably linked to sugarcane and rum since the colonial period, when the Dutch were establishing trade routes in the West Indies. During the eighteenth century, John Adams had the temerity to suggest that the American Revolution was really about one item—molasses, which was the main ingredient used in rum and was defined by the British as sugar.[1] As pioneers moved west, the contradiction of allowing slavocracies in a democracy became much more clear when territories organized into states. Although I don't want to simplify the causes of the Civil War, I want to suggest that the sugarcane industry was a catalyst because sugar was everything honey was not—cheap, convenient, and free from stings—and slavery was tolerated in order to have it.

African slaves labored on sugarcane, indigo, and rice plantations long before cotton was an established industry in the southern states. Because sugarcane could spoil rapidly, slaves often worked eighteen to twenty-hour days during harvest so the crop would not be lost. Given the stressful elements of the tropics (malaria, snakes, and heat), the average life span for a male slave on rice and sugar plantations was no longer than thirty years. As slaves were gradually exposed to, and accepting of, Christianity, they used the Bible to find hope and freedom.

Yet as slaves gradually accepted the Christian religion, they used the Bible to find hope and freedom. They readily identified with the stories of Israelites and their search for Canaan, the land of milk and honey. In their songs, slaves appropriated references to Canaan. Ethnologists recorded slave songs after the Civil War. Of the south Georgia spirituals, "The Lonesome Valley" was one of the earliest recorded by Allen, Ware, and Garrison in 1867.[2]

> My brudder want to get religion? Go down in the lonesome valley.
> Go down to the lonesome valley to meet my Jesus there.
> O feed on milk and honey. Go down to the lonesome valley.[3]

The slaves focus on Canaan as a promise for deliverance from slavery, leaving free the possibility to wonder if Canaan is the North or simply an emotional plateau.

However, the slaves also could take a spiritual and turn it into a satire, as illustrated in the case of an old English folk song:

> The Lord made the bees,
> The bees made the honey,
> The Lord made man
> And Man made money.

In 1863, Frances Anne Kemble recorded the same verse with a twist from a Negro boatman's song while visiting in Georgia:

> God makes the bees
> The bees makes the honey.

God makes man,
Man makes money.

By the time Joel Chandler Harris heard the song in the 1880s, blacks had turned it into a sharp satire about the capitalistic ways of whites:

De old bee make de honeycomb
De young bee make de honey,
De n— make de cotton an corn
An' de white folks gits de money.[4]

Harris recorded this song and other proverbs in *Uncle Remus,* which appeared after the Civil War. *Uncle Remus* reflected and recorded how slaves reversed the value system in which they labored. The slaves—not the white masters—were the industrious ones, and the whites were lazy. Other sayings worth noting: "It's a mighty po' bee dat don't make mo' honey dan he want" emphasizes industry. Another proverb, "Hit takes a bee fer ter git de sweetness out'n de hoar-houn' blossom," emphasizes the importance of maintaining an optimistic outlook when dealing with life's bitterness.[5]

Recipes showed that blacks from honey producing regions in Africa remembered food patterns and incorporated them into the plantation menus. According to a slave narrative taken from Shad Hall in Sapelo Island, the writer's grandmother would make a dish called *sakara.*

> She make strange cake, fus ub ebry munt. . . . She make it out uh mean an honey. She put mean in bilin watuh an take it right out. Den she mix it wid honey, and make it in flat cakes. Sometime she make it out uh rice.
>
> Duh cake made, she call us all in an deah she hab great big fannuh full an she gib us each cake. Den we all stands roun' table, and she says, "Ameen, Ameen, Ameen," an we all eats cake.[6]

Recipes and songs are the few texts that provide proof that the slaves had used honey in Africa before coming to America. Once established in America, the slaves appropriated the biblical image to refer to the North, to the afterlife, or to remember their past African heritage.

As soon as South Carolina troops fired on Fort Sumter in 1861, honey rose in value. The sugarcane industry in the South was suspended, not just for the next four years but for fifteen. Southern lands were ravaged, and southern railroads were destroyed. The slave economy was in turmoil. More people (Northerners and Southerners) were affected by the loss of sweeteners than at any time before the colonists' arrival in the 1630s. The Civil War affected beekeepers in negative ways. The *American Bee Journal,* started in 1861 by Samuel Wagner (Langstroth's good friend) in Philadelphia, was suspended during the war.[7] *ABJ* informed many people how to be beekeepers before schools taught the science and gave people a venue to talk about common problems.

Many beekeepers went to war. Lorenzo Langstroth's own son fought for the Confederacy even though his French Huguenot ancestors abhorred slavery. Others fought in the Army of the Grand Republic. Capt. J. E. Hetherington of Cherry Valley, New York, a successful beekeeper before the war, was wounded three times. During one battle, "his sword was struck and bent by a bullet that would have pierced the Captain's heart while he was in the thick of fighting."[8] He returned to his hives after the war, although "his life remained in jeopardy for two years," according to historian Wyatt Mangum.[9] Another soldier, O. O. Poppleton, entered the Iowa troops as a private and worked up to the rank of lieutenant and colonel.[10] Dr. Godfrey Bohrer, from Ohio and Kansas, worked as a surgeon for the Northern troops.

Honey-hunting skills were important to both armies. In the South, soldiers were on leaner diets than the Northern army. So when Gen. Joseph Johnston's army was in Mississippi, a region known for its excellent honey, the soldiers took advantage of their good fortune. In an article written extolling military life in the *Louisville Courier-Journal,* the writer records a bee-hunting episode:

> When Johnston's army was about twelve miles from Vicksburg, orders were issued that death would be the portion of any man who fired a gun, chopped with an ax or made any noise whatever. . . . One day some soldiers noticed that some wild bees had selected an old dead tree as the depository for their honey. Away up in the top the little workers could be seen storing their sweets.

The soldiers eventually talked a countryman into setting fire to the tree, and when it crashed, the countryman was nowhere to be found. Hence

4.1. "Gone Off with the Yanks." Courtesy of the Library of Congress. Edwin Forbes (1839–1895) drew this cartoon as part of a larger series titled, "Life Studies of the Great Army." This soldier's activities are known among modern beekeepers as "the unknown transfer of ownership of hives."

no one was punished. The bees were smoked out, and the soldiers feasted on the honey "like dogs upon a lame coon."[11]

But Union soldiers were equally skilled at foraging for honey. While writing home from Mississippi, Henry Schafer of the 103rd Illinois gave a detailed list of confiscated items: "Last night was a jolly time in camp with some of the boys as hogs, calves, sugar, honey mollases [*sic*]. came into camp plentifully."[12]

Of course, some soldiers didn't know what to do with bees, and those occasions provided a few humorous instances during these dark times. A cartoon shows the joys of stealing bee skeps. Hattie Brunson, writing in her diary as a young girl in South Carolina, recalled seeing one soldier from Sherman's army attempting to steal honey just as the rest of his peers were robbing their plantation home. "His face was just black with bees!" she recalled gleefully.[13]

The 132nd Pennsylvania infantry regiment found itself in a tight spot at the battle of Antietam when it advanced upon William Roulette's farmyard on September 17, 1862. Roulette was a beekeeper, and his bees did not take kindly to having their hives overturned by the crushing movement of Union soldiers. In the midst of gunshot, smoke, and hot temperatures, the bees became another enemy. "With bullets and artillery shells whizzing through the putrid air," writes Kent Masterson Brown, "thousands of swarming bees created an impossible situation."[14]

Roulette's bees were not the only ones disturbed by the Civil War. Two years later, when the war was all but decided, honey bees in Okolona, Arkansas, showed a truly nonpartisan dislike for both Union and Confederate soldiers. On April 3, 1864, Gen. Joseph Shelby's Confederate troops clashed with Gen. Samuel Rice's Union troops. During a severe thunderstorm (complete with hail and high winds), several beehives were overturned. Although the bees first attacked the Confederates, they were just as furious in attacking the Union army. Both sides decided to call it quits and resumed fighting the next day.[15]

Perhaps suggesting the bees ability to defend itself, the 72nd Pennsylvania Fire Zouave regiment adopted the honey bee as a mascot to fly on its left General Guide Marker. Historian Jason Wilson explains, "This would not be the official flag for the regiment, as they had another state-issued color that was standard for all Pennsylvania infantry regiments. This particular guidon was originally owned by Lieutenant Thomas F. Longaker, of Company F, and was donated to the state's collection in 1930."[16] In contrast to the billowy, scarlet Zouave uniforms, the flag itself is more subdued, having a blue field, with a white oval; a golden honey bee is in the middle. The Pennsylvania Fire Zouave unit, the showiest of the Union troops, was known for its brilliant uniforms, precise marching maneuvers, and steadiness in combat, and perhaps for these reasons, some independent soul decided to create a flag with a mascot that was known for stability.

On the business side of the Civil War, Moses Quinby was getting top dollar for his honey in New York. During his lifetime, Quinby owned or partnered in as many as 1,200 hives. But it was during 1865, the last year of the Civil War, that Quinby shipped eleven tons of honey to New York City, "causing both quite a stir in the press and a strain in the market."[17] Because of his pacifist religious stance, he stayed out of the war but profited handsomely.

The Shakers in Pleasant Hill and South Union, Kentucky, also pacifists, did not profit nearly as well because they were located in a border state. The Shakers emptied their stores to feed both Northern and South-

4.2. Seventy-second Pennsylvania Flag. Photo courtesy of the Pennsylvania Capitol Preservation Committee, Harrisburg, Pennsylvania. Lieutenant Thomas F. Longaker, of Company F, owned this flag and used on the left guidon; it was donated to the state's collection in 1930.

ern troops passing through. They did not charge for their meals, nor did they participate in the war. In their communities, all people were equal and supposed to work.

In spite of the stresses caused by the Civil War and Reconstruction, the Shakers used the honey bee in hymns to lift the morale of their members. At the Shaker community in Lebanon, New York, "Busy Bee" was just one of the songs Shaker Anna White "received" in a spiritual vision:

> Like the little busy bee,
> I'll gather sweets continually
> From the life giving lovely flowers, which beautify Zion's bowers.
> No idle drone within her hive, will ever prosper, ever thrive,
> Then seeds of industry I'll sow, that I may reap wherever I go.[18]

This hymn and other hymns collected by beekeeper Elder Henry C. Blinn in *A Sacred Repository of Anthems and Hymns* (1852) encour-

4.3. Elder Henry C. Blinn of Canterbury, New Hampshire, Shaker Village. Courtesy of Hancock Shaker Village. A former teacher, Blinn was a good beekeeper and recorded many Shaker hymns using honey bees as a symbol of heaven on earth.

aged Shakers to see themselves as bees able to serve a divine purpose.[19] Blinn, a beekeeper until the early 1900s, "seemed to be very fair of his judgment" by all accounts, states Pleasant Hill Shaker historian Brian Thompson. Given that Blinn had been a teacher before becoming a Shaker, he was "very observant," according to Deborah Burns.[20]

It should be noted that bees helped to serve a financial purpose as well. The Shaker brethren were among the first Kentuckians to experiment with Italian bees once they become available.[21] They were also some of the first queen breeders in Kentucky. Although the Shaker culture had very little iconography, the bee was one of the few consistent Shaker symbols. Although the Shakers survived the Civil War into the twentieth century, the communities could not keep up with the multifaceted changes taking place after the war.

The most poignant Civil War exchange involving a beekeeper occurs in letters written by John W. Wade to his wife Rutha Cox Wade. Drafted to serve for the Confederate troops in 1862, Wade was in his thirties, had a small farm, and two children, with a third on the way. He fought in some of the worst battles of the Civil War. As he continued to survive each one, he dreamed of returning home. In a letter dated January 14, 1864, he admitted sadly, "A few nights ago I dreamed of taking honey out of the bee hives and eating of it. I study and dream a good deal of home, so much so it makes me uneasy."[22] Wade died at Spotsylvania a few months later.

After the Civil War was over, beekeeping as a hobby and industry escalated in the North. Railroads expanded into the western territories. The wartime industries changed back to domestic goods. Printing presses were back in business, publishing *American Bee Journal* and *Gleanings in Bee Culture: Or how to Realize the Most Money with the Smallest Expenditure of Capital and Labor in the Care of Bees, Rationally Considered.*[23] A. I. Root and his magazine (now titled *Bee Culture)* established a unified presence in beekeeping because they were in one central location, whereas the *American Bee Journal* changed editors and places of publication several times before it finally found a home in Hamilton, Illinois.

A. I. Root transformed the image of the beekeepers in America. He was not James Fenimore Cooper's bee hunter: semicivilized, wild, and living on the frontier. In fact, before he became a beekeeper, he was a jeweler living on Main Street and settling down into Christian ways. One day, while looking onto the Medina, Ohio, square, Root spied a swarm fly into a tree. He made a friendly bet with his employee that basically offered the employee a dollar if the man could woo them down. The man caught the swarm, collected his dollar, and convinced Root that bees were a fascinating wonder.

In short order, Root read Langstroth's book and struck up a friendship with the pastor. According to historian Pellet, Root did the most to commercialize beekeeping by standardizing the Langstroth hive in 1869.[24] He also published a how-to book called the *ABC and XYZ of Bee Culture,* which arranged topics alphabetically and remains in print. The

Your old friend,
A. I. Root.

4.4. A. I. Root. Courtesy of *Bee Culture*. The friendly caption at the bottom of this photo marked a transition in the public's perception of beekeepers. A businessman, Root adapted nineteenth-century industial principles to beekeeping. His magazine, now known as *Bee Culture,* provided an early forum for beekeepers.

straightforward friendly tone of *Bee Culture* was an immediate success with readers. In its inaugural issue, Root states, "Improved Bee Culture is our end and aim, and we trust no one will hesitate to give any fact from experience."[25]

By the end of the century, Root formed one of the major bee supply companies in the world. It had a bee yard, a railroad spur, its own water tank, a printing press, and a lumberyard. In keeping with the progressive nature of social reform, Root would help his employees if they needed money. But he could be ruthless with dishonest people, publishing a "Swindler's Column," which provided a checks-and-balances system to protect his customers. In one particularly damning column, Root took George Moon to task for producing an unprofessional magazine, *Bee World,* in 1874: "If the man can neither read, write, nor spell himself, he certainly should not leave his readers to infer that no one in Rome, Georgia, can do any better . . . we have never before seen any thing so lamentably deficient in the principles which any common school education should give, as Mr. M's attempts at editorials."[26]

In the same column, Root praised *Mrs. Tupper's Journal,* in which he says that Mrs. Ellen Tupper "presents us one of the most valuable nos. yet issued." In fact, he invited women to send letters. In one response to Cynla Linswick, who worried that her letter would be too long to include in the issue, he assured her: "Please do not imagine that articles like the above are in any danger of being found too long for [*Bee Culture*]. . . if we have so many letters from our lady apiarists that we cannot find room, we shall enlarge [*Bee Culture*], we assure you."[27] It is hard to overestimate Root's importance at galvanizing an expanding beekeeping community, which had splintered during the Civil War.

The South had difficult adjustments to make. Soldiers returned to find their fields and homes destroyed; railroads, likewise. Industry was nonexistent. For many white soldiers, it was easier to move to the West than remain in the South. The 1862 Homestead Act promised free land if settlers could survive on it, and many decided to move, taking their beehives with them. Kansas historian Edward Goodell describes in great detail how skeps were brought to the West:

> A light four-wheeled cart was built especially to carry the skeps. . . . A framework of hickory poles was built over the car, and over these a strip of canvass [*sic*] could be pulled in bad weather; in fair weather this was taken off, and fresh cut green boughs were laid across to provide shade for the bees. A long detachable double length tongue with a metal ring in the end of it was used to fasten the cart to the rear of the covered wagon in which the family traveled. . . . The purpose of that long tongue to the bee-cart was of course to keep it as far as possible from the covered wagon pulling it.[28]

Those who did not carry their hives often found bees waiting for them in Texas, Missouri, Illinois, and Kansas.

One Texan, Edouard Naegelin Sr., decided to start a bakery in New Braunfels, given that he had a steady supply of honey. Arriving from Alsace-Lorraine, France, in 1844, he fought on the Confederate side when he was nineteen. Four years after the war ended, Naegelin returned to New Braunfels with a sack of flour and less than a dollar in his pocket. In short order, he bought the original baker's lot belonging to Heinrich Zuschlag, the baker to Prince Solms. Then Naegelin advertised in the paper that he was "baking all kinds of bread and cookies."[29] He was especially noted for making *lebkuchen,* a light cookie made with honey.

Perhaps the most neglected honey bee researcher was born two years after the Civil War.[30] Charles Henry Turner, an African American, was born two years after the Civil War in Cincinnati, Ohio. "Turner was a pioneer of the comparative psychology/animal behavior movement in America," according to scholar Charles Abramson. Turner's experiments suggested that honey bees could see colors and discriminate among them. Following his initial experiments with color he performed more sophisticated experiments with patterns by suggesting that honey bees could be conditioned to fly to certain stimuli, provided they were trained. In some ways, his studies prefigured those of Karl von Frisch, although Frisch never acknowledged Turner. Even though his greatest achievements would occur in the twentieth century, Turner was preparing for the challenges in admirable fashion: he was selected valedictorian of his high school class and became the first African American granted a doctorate from the University of Chicago.

Although the sugar trade was disrupted by the Civil War, the Appalachians did not suffer from the lack of sweeteners. Timber made it possible for people to have bee gums, even though the honey was often likely to be raided by armies from both sides. One mark of how many bee swarms existed is an article which shows that, in the Smoky Mountains' Cataloochee section, two thousands pounds of honey were taken in the late 1860s.[31]

Fortunately, the Civil War did not discourage Charles Dadant from immigrating to Hamilton, Illinois, in 1863 with dreams of starting a vineyard.[32] At age forty-six, Dadant arrived penniless after paying for his family to come to America and buying a farm in Hamilton. He also did not know a word of English, but he was determined to succeed in this country. Even though there were French-speaking people in the area, because some Icarians had stayed in Nauvoo, Dadant learned English by subscribing to the *New York Tribune.*

Dadant turned to beekeeping, a hobby he learned while a clerk in Langres, France, an area known for its dry goods business in locks and knives, when his dreams of being a vinter were not fulfilled. According to biographer Kent Pellett, Dadant educated himself while a traveling salesman throughout the French countryside. As the horse would plod

4.5. Charles Dadant and John Hammon. Courtesy of Dadant Company. A French socialist, Dadant believed in working in the factory along with his workers. Exodusters like John Hammon, women, and immigrants enjoyed progressive benefits at his factory. Dadant set up savings programs and shared profits with his employees. When he became editor of *American Bee Journal,* he provided an international forum to the beekeeping community.

along a well-traveled road, Dadant would read the works of French biologist Lamarck and the theories of socialist Charles Fourier. Dadant, always bothered by the ever-present poverty in his region, renounced the Catholic Church and became a socialist.[33]

When he moved to America, he modeled the early stages of his business on Fourier's principles. Even as the business grew, Dadant worked alongside his employees, sharing in the difficult stages of many tasks. According to Pellet, Dadant hired people of different races, ethnicities, and genders: "carpenters who made hives and packing boxes, men for dipping foundation, men who worked about the farm, and during vacation days, girls who interleaved the sheets of foundation."[34] John Hammon, a former slave, is mentioned twice in Kent Pellet's biography of Charles Dadant. Both episodes illustrate the philosophies that Dadant had when it came to caring for his employees. Hammon initially had difficulty handling moncy, something that Dadant could also appreciate because his son Camille Pierre was the financial genius behind the company's success. Charles Dadant initiated a savings program for Hammon so the former slave could buy land. The second occurrence was when Hammon finally located his mother in Georgia. Dadant himself happily broadcast the news to everyone in town and wrote letters to the *Review Internationale* and American papers.[35]

In addition, Dadant paid his employees by piecework, which cost more: the employees "did their work more promptly, which was very desirable, as customers were always in a hurry for their orders."[36] Dadant was "delighted" when an employee would throw out an imperfect piece of wax comb foundation. "Visitors at the little factory were surprised at the atmosphere of cooperation and at the zeal of the workers," concluded Pellett.

The same type of good will characterized Dadant's friendship with Lorenzo Langstroth, who could speak French.[37] Although there were religious differences, their friendship surpassed those.[38] In fact, Dadant first described the Langstroth hive to French beekeepers and served as an unofficial liaison between the two countries until his death. Langstroth succeeded in importing the Italian bee into the United States before the Civil War, and with Ellen Tupper's help, Dadant secured funding for future imports when foreign trade resumed. In fact, it was he who figured out an easier way to ship bees across the Atlantic. Dadant even created his own hive, which had success in France and America. A prolific writer, he championed Langstroth's hive against those who claimed that Langstroth wasn't the inventor. Having had marked experience with the log hives and skeps in France, Dadant knew firsthand Langstroth's contribution and said so in the *American Bee Journal*.[39]

More importantly, Dadant took over the printing and publication costs associated with Langstroth's *The Hive and the Honey Bee*. Because he suffered from debilitating headaches, Langstroth could not bear the burdens of revisions and updates. The Dadants offered to publish the revised version and translated his books into French, Italian, Polish, and even Russian. Therefore, Langstroth's work was known around the world within a very short time. When the *American Bee Journal* needed a new editor, Dadant was a natural choice. Although his English-speaking skills were always awkward, his prose was quite solid, and he had a wide network of friends and colleagues.[40]

The Civil War exacted a physical and mental toll on many veterans, including beekeepers such as Langstroth's son; Capt. J. E. Hetherington; and honey hunters headed West. There is a sense of military strategy added to the honey-hunting tales that didn't exist before the Civil War. And perhaps the hard feelings between the two regions lingered longer than they should have, which could have hampered Texas beekeeping, according to bee historian Clark Griffith Dumas. In his thesis, Dumas argues that Texans did not appreciate having black soldiers impose civil curfews during Reconstruction. It was a credit to beekeeper Judge Andrew McKinney that he could keep order and encourage beekeeping.[41] Dumas also speculated that people in the South were too poor to be able to travel the long distances to beekeeping conventions, often held in Toronto, Cincinnati, or Chicago.[42]

Fortunately, at the 1906 National Beekeeper's Convention in San Antonio, Dr. Godfrey Bohrer received a bouquet for being the oldest member of the National Beekeeper's Convention. He used the occasion to mention the stark contrast of the harsh feelings between the North and the South and the present brotherly feeling of being a Union. The speech, according to the writer, was so full of feeling that it "brought tears to the eyes of several old Confederates still present."[43]

1870 to 1880

The decade begins with an emphasis on educating women about beekeeping. Although women had been *pictured* with bee skeps, few written records existed about women beekeepers before the nineteenth century. Eva Crane notes that "most women who did any bee work assisted their husbands," but she is not specific about the time periods.[44] Scholar Frederick R. Prete, writing about British beekeeping textbooks, convincingly argues that by the Restoration "gender distinctions [in England] became more ossified, and gender roles were seen to be separated by nature as well as by function."[45] The British bee books were written to convince women beekeepers that they should be chaste and submissive, even though the biology of the hive argued quite the opposite.

In America, however, technological improvements made beekeeping easier, but women had to depend on themselves as a result of the Civil War, and the gender differences became blurred. Women participated in the beekeeping community by writing to American bee journals, attending conventions, and keeping bees. Women became members in beekeeping societies, and several even edited beekeeping magazines and owned their own apiaries.[46] In fact, the decade began by promoting the interests of women beekeepers: "It is interesting to note that the National Beekeeping Federation constitution adopted in 1871 provided that ladies might be admitted to membership without charge, while the fee for men was one dollar each," muses Frank Pellet.[47]

When the *American Bee Journal* resumed publication after the Civil War, Mrs. Ellen S. Tupper of Brighton, Iowa, garnered national attention as a frequent contributor, editor, and teacher. Tupper was a well-known beekeeper, queen rearer, and editor of the *National Bee Journal*.[48] Her peer, A. I. Root, praised her work: "We shall have to conclude that woman's taste is certainly equal, if not superior to that of the sterner sex, in such matters. The typography and general appearance externally, and the whole work certainly does her credit."[49]

Frank Pellet briefly mentions an article published in *The Illustrated Beekeeper's Journal,* which published Tupper's plan for queen fertilization in 1870. The writer, L. C. Waite, testified that Tupper's plan worked, but nothing more is mentioned about using her wire cage method.[50] Nonetheless, Tupper regularly published after 1865 and even taught at Iowa State College in the 1880s.

Charles Dadant's trip to Italy in 1872 was to import large shipments of Italian queens to his and Tupper's apiaries. They each wanted approximately a hundred Italian queens, and Dadant thought if he could pack and unpack them, he would learn how to import queens.[51] Through trial and error, Dadant finally realized how to ship bees so that they would arrive in good condition and established a relationship with a shipper who would follow his directions. Tupper, however, "had abandoned importing [Italian bees] after she had lost several thousand dollars."[52]

There were other female beekeepers. At one convention, for example, a lady speaker, not named in the meeting's notes, talked about the benefits of beekeeping for women, saying, "It was a very pleasing occupation, and about the only outdoor one for women to pursue."[53] Another oft-mentioned beekeeper was Anna Saunders, who lived in Mississippi. In 1874, she specified geographical differences that affect beekeepers: "We must have a journal on Apiculture in the South. Our wants are so different from yours; our troubles are chiefly summering, over swarming, and insects of whose annoyance you can scarcely form a conception there."[54]

Yet there were obstacles for women to overcome. Cyula Linswik complained: "One of the thorns in the path of the woman who undertakes to master the theory and practice of beekeeping is her lack of natural or acquired ability to drive a nail straight, to use a saw with safety to the implement, or a sharp knife with safety to herself [. . .] And the woman who begins to keep bees without having her attention directed to this matter is in danger of suffering from vexation of spirit and wounded fingers many times during the course of her novitiate."[55] Linswik questioned the benefits for women. "Does apiculture offer any special inducement to women? May it not be that the work, no longer impossible, is still for them undesirable?"[56] In fact, Linswick shared a poem written to her by a neglected friend. The writer's humor doesn't mask the real loneliness that women must have endured in the 1870s:

My eyes have grown big,
And my ears have grown long,
Watching and listening for you;
Every morning I say,
"She will be here today!"
But the prophecy never comes true.
Then I think of your bees,
Round the dry, leafless trees,
And I think, as I fancy them, humming,
That the day may be warm,
And the rascals will swarm,
Just enough to prevent you from coming.

The writer signed the poem, "Yours crossly!!"[57] Linswick concurred that the bees were the reason why she had not visited her friend, and in her letter to the editor, she promised that she would make amends.

Other women became equally prominent beekeepers in the South. Sarah Elizabeth Sherman was born in Georgia to an impoverished family. When she made an alum basket and filled it with wax flowers, her father sold it for seven dollars. From that moment, she "began to feel that she was not a cipher in the arena of life, but a living reality, and that she could do something to help make a living."[58]

She moved with her family to Texas, married, had a son, and settled down. Widowed at a young age, she began to rent a section of her farm and keep bees on the other section. Starting with black bees in the 1870s, she quickly switched to Italians as soon as she could. In spite of the frontier conditions, she kept her own hives, constructed them, sold them, and even exhibited her honey at the World Exposition in Paris, France.[59] She was a frequent and forthright contributor to *Texas Farm and Ranch*.

Jennie Atchley, who lived in Beeville, Texas, had between eight hundred and one thousand colonies devoted exclusively to queen rearing. Born in Tennessee, she moved to Texas in 1876 and began beekeeping. The Atchley family kept Jersey cows, hogs, and chickens. Atchley was a transitional figure between nineteenth-century beekeeping and twentieth-century beekeepers. She briefly published *The Southland Queen* at the turn of the century, but eventually moved to California.[60]

Even in the Hutterite communities, which migrated to the Dakota Territory in the 1870s, equality between men and women was promoted in an effort to share everything.[61] During the nineteenth century, men and women shared the work associated with beekeeping and honey production, but the chores broke down along gender divisions. The social role of beekeeper, for example, was generally given to a man, although women often did most of the honey production.[62]

Mormon women beekeepers were given as much respect as men in their society. In 1885, Edward Stevenson wrote to the *Deseret News*, "Brother George Bailey of Mill Creek told me last week that he had already taken 5000 pounds of honey with the aid of his son and wife, who by the way is very efficient and more than equal in the business to most men."[63]

Ironically, given its emphasis on equal roles between genders, the

4.6. Woman beekeeper. Courtesy of *Bee Culture*. Date unknown. Identity unknown. Women wore long dresses in the field until the twentieth century, although Moses Quinby argued in 1879 that "every woman should wear a dress suitably short. . . . drawers should be tolerably wide, gathered on a band at the bottom, and buttoned tight about the ankle."

Shaker community at Pleasant Hill, Kentucky, relegated this task to the men. Women canned and put up preserves, while the brothers took the honey. In 1872, according to Julia Neal, the Shaker sisters "finished 155 quarts of cherry preserves and attended two meetings. The same week, some of the brethren took 212 pounds of honey."[64]

Women in California were making a presence in beekeeping. Root encouraged Dr. J. J. Jansco and especially his wife not to be depressed by the labor-intensive nature of beekeeping. Root wrote in *Bee Culture:* "Remember that a nation of sisters are debating whether they are fitted for such duties . . . and even one who gets discouraged and gives up may exert a wide influence over the rest. . . . remember what a great boon it will be to many of your sex if they once learn that they can thus be useful, and feel that their acquired skill and knowledge places them where they may not feel dependent on others, no matter what reverses may overtake them in life."[65] Wise advice.

Women weren't the only ones going to California. By 1878, California had abundant sources of nectar and the bees to harvest them. This triggered another gold rush, except this time the gold was honey. But beekeepers lost their stakes just as the forty-niners did if they weren't mindful of their circumstances. An unfortunate example of a beekeeper who went west with dreams and ambitions was Rufus Morgan.

A photographer and beekeeper in North Carolina, Morgan was convinced that he could make a healthy profit in California. He left his pregnant wife Mary Clark Morgan behind while he went to establish a place in San Diego. Upon arriving, he struck up a business relationship with successful bank director Ephraim W. Morse, in which profit and expenses were to be shared equally if Morgan would take charge of Glen Oak Apiary. In a series of fifty-five letters to Mary, we see Morgan become more realistic about the honey conditions and his business relationship with Morse. Of course, Morgan also expresses his longing to be reunited with his family and his dreams of having them in California.

The letters from January 1879 reflect Morgan's initial optimism: "The honey business is a big thing here, bigger than I had any idea of—I *know* in five years from now I will be making $5000 pr year!—Fortunes are made at it here & *all* are doing well at it. Plain farmers talk of raising $1500 to $2000 worth of honey just as we talk of making 8 & 10 bales of cotton" [emphasis Morgan's]. But letters dated in February suggest that Morgan's dreams would be in trouble, for there was a "dearth of pollen," and in a letter dated March 12, 1879, he admits that "everything is higher here than there." On March 28, "Everyone has the blues" because there had not been any rain.

By May 1, 1879, Morgan verifies that the weather did not cooperate with his plans, but he and Morse were negotiating the purchase of three more apiaries. Still, he was running up debt with Morse and not bringing in as much money as he thought. Morgan began to keep chickens and make wooden decorations for sale as he waited for a turn in fortune. In a letter that contained an ominous prediction, Morgan happily predicts, "If [Morse] does not die, I will have *all* the capital I *want*."[66]

Unfortunately, a year later Morgan died unexpectedly after eating a meal of poisonous mushrooms. His dreams of becoming another Samuel Harbison never materialized. In fact, according to transcribers Massengill and Tompkins, Morgan's partner Morse sent a letter to Morgan's wife informing her that Morgan owed him $125 dollars and offered virtually nothing in the way of compensation. Morse did return Morgan's earthy possessions—a gold watch, some gold studs, some pressed ferns, and a few books—but offered very little in the way of condolences. Barbara Newton provides several clues to Morse's financial circumstances, which were precarious due to the drought—Morse had lost several thousand dollars by that point. Morgan died in April 1880, so Morse had to find someone to take over the apiaries quickly because the colonies were growing, and nectar would soon be available. It is possible that he did not have time for the social niceties which sad occasions demand. Fortunately, Morse stayed in beekeeping. He unified beekeepers in terms of marketing and shipping honey overseas because the Central Pacific railroad had a monopoly in the region and charged exorbitant shipping rates.[67]

Morgan was not the only person to migrate from the South. When Union troops deserted the South in 1877, Exodusters (former slaves) left for Kansas, Missouri, and Illinois. As already mentioned, one Exoduster, John Hammon, stayed in Hamilton, Illinois, to work in Charles Dadant's fledgling bee supply industry. Too, a small contingent of blacks went to Utah with the Mormons, often serving as cooks and hired hands to the families that had left the South. Thus, Reconstruction was a profound time for American beekeepers who, along with the larger society, often rebuilt new lives in new regions with new ethnic groups.

Inventions

The four major inventions transforming beekeeping from a sideline interest to an industry appeared within a fifteen-year span: the moveable frame hive (1851), the wax comb foundation (1857), the centrifugal honey extractor (1865), and the bellows smoker (1873). Langstroth's moveable frame hive was the prototype for many new hives in America, but they all adhered to the same principle of bee space.

In moveable frame hives, bees build their comb on sheets of foundation—a thin sheet of beeswax embossed with hexagonal cell walls—placed vertically in the middle of a frame. This foundation guides the

4.7. Johannes Mehring. Courtesy of Dadant Company. This man invented the comb wax foundation maker, the second invention needed for a commercial beekeeping industry.

bees when they build their cells and respects bee space between frames and the hive body. Johannes Mehring invented a machine that would manufacture embossed wax foundations in 1857. Using Mehring's ideas, A. I. Root developed a similar roller press in 1876. His first customer was Christian Diehnelt.[68] Eric Nelson sums up, "Further development of wax foundation has been a process of refinement rather than innovation."[69]

Next came the extractor, which took honey from the wax foundation sheets. Before the extractor, people had no alternative but to remove honey by "cutting the comb from the frames, crushing it, and straining the honey from the wax. This was both time consuming and costly," according to Nelson.[70] Using the idea of centrifugal force, Austrian Fransesco de Hruschka created an extractor. After removing the wax cell cappings from a frame of honey, he placed the frame in a basket

4.8. Two workers pressing comb foundation. Courtesy of *Bee Culture.*

located in a drum-like canister. Centrifugal force pulls the honey out of the cells when the basket is turned, and honey collects at the bottom of the canister.

In Boston, Massachusetts, H. O. Peabody created an extractor that had baskets balanced on a pivot arm in 1870. When T. W. Cowan improved on this by making the baskets automatically reversible so the frames did not need to be turned by hand, the stage was set for a less labor-intensive chore than honey extraction had been initially. "The extractor made liquid honey available on a commercial scale previously unknown," according to Nelson. "However, many people refused to believe that it was pure honey, and several years passed before it was accepted by the public."[71]

In 1873, Moses Quinby improved the smoker by adding bellows to a firebox. Before, beekeepers had difficulty keeping the smokers working. Quinby's model increased the airflow in the firebox, greatly im-

4.9. Major Fransesco de Hruschka. Courtesy of Dadant Company. This man invented the honey extractor. The invention was the third step in forming a modern bee industry. A popular beekeepers legend is that Hruschka saw his child swinging a basket over his head and that action inspired him to develop a similar basket to hold honey frames.

proving the life of the coals inside. With the bellows attached, however, the smoker could be used with more control on the beekeeper's part. "At last, the fundamental principle was found," writes Pellet. "Once well fired, it [the smoker] worked beautifully, but it went out when laid aside."[72] Having made the invention at the end of his life, Quinby never tried to patent it. He died shortly thereafter. These four inventions—essentially unchanged—remain the pillars that support the beekeeping industry, although modifications have been made through the years.

In addition to these discoveries, queen rearing had largely been handled by European scientists until the 1870s. The major principle concerning queen rearing—that is, that worker bees can produce queens from young larvae originally in worker cells—had been discovered as early as 1568 by Michael Jacobs, but his observations had been met with disapproval.[73] According to Harry Laidlaw Jr., "The discovery that larvae in worker cells can develop into queens is as important to queen rearing as Langstroth's recognition of the bee space is to beekeeping as a whole."[74]

4.10. Honey extractor. Courtesy of *Bee Culture.* The extractor uses centrifugal force. The frames are placed in baskets and then turned with the crank. Honey slides down the walls and collects at the bottom. A valve opens to let the honey drain into bottles.

Nineteenth-century European scientists Francois Huber, Johann Dzierzon, Miss Jurine, Karl von Siebold, and Rudolph Leuckart were able to prove Jacobs's assertions, and their discoveries began to encourage those who wanted to study bee genetics. Gregor Mendel's work with sweet peas was just beginning to unlock the secrets of inheritance. But many of the early experiments with queen bees were hampered by beliefs that queens mated in the hive.

Still, when American beekeepers such as Langstroth, Samuel Wagner, and Moses Quinby learned of the Italian bees, they wanted the subspecies to be available. Importing Italian queens fueled interest in queen rearing and the search for the perfect bee. Although early experiments in queen rearing were not successful, a beekeeper named G. M. Doolittle "showed great ingenuity in combining the advantages of numerous meth-

4.11. Quinby smoker. Courtesy of Wyatt Mangum. By attaching a bellows system to a smoker, Quinby invented a fourth tool for commercial beekeeping.

ods advocated by others into a workable system."[75] He made artificial cell cups and transferred young worker larvae into them. By placing a small amount of royal jelly in these cups (known as "grafting"), Doolittle increased the chances that the cups would be accepted by the bees. Using a suggestion by A. I. Root, Doolittle dipped the cells in warm wax and did his grafting in a warm room. Again, he met with success. Finally, he helped develop a queenright cell builder colonies and cell finisher colonies, which have become standard in the beekeeping industry.[76]

Rolling on the River

Migratory beekeeping became possible during the 1870s because trade routes suspended during the war were reestablished, some by unconventional methods. C. O. Perrine wanted to have migratory beekeeping on the Mississippi River. He invested in a fleet of barges, hired fifteen men, and collected one thousand colonies of bees. Even by today's standards, that dream would have seemed ambitious, but Perrine liked large dreams. "He had visions of shipping the honey direct to Europe, as he was exporting extensively at the time."[77]

His dream did not make it through the season. No one connected with the enterprise had a clear sense of honey flows and blooming seasons. When the barge should have anchored near some banks for a week or two, the captain would move in one or two days. Furthermore, so many machinery problems occurred so often that Perrine gave up on the barges and loaded the hives on a steamer. The operation made it as far north as St. Paul, Minnesota, and the bees were taken south again, to winter on a bank in Calhoun County, Illinois. When yellow fever subsided in New Orleans, Louisiana, the bees finally floated home.

Perrine suffered heavy losses, as much from hives falling in the river as from bees flying away from his barges, to say nothing of poor honey production. His last words on floating apiaries were spoken at the North American Beekeeper's Association, when he was reported to have advised beekeepers: "Keep as far as possible from large bodies of water."[78]

Later, O. O. Poppleton transferred his business from Iowa to the Indian River region in central Florida. Because he chose hives that were not top-heavy, Poppleton could move his hives up and down the river at will, depending on the bloom season, proving migratory beekeeping could be done on waterways.

During this time, railroads were the best choice for migratory beekeepers. Although railmen initially resisted the thought of carrying bees, many beekeepers profited from the large, open cars, and they were able to get two seasons from their bees. A particularly colorful character is a man named Migratory Graham, who at one time had three thousand colonies in California. He produced 120 tons of honey in one season, but he was perpetually in trouble because of the disease laws he violated with his large colonies.

Still, a horse and wagon could carry bees just as well, and in 1872, Gen. J. B. Allen brought bees from San Diego, California, to Tucson, Arizona. On July 27, 1872, the *Arizona Citizen* happily predicted: "The bees brought here from California by Gen. J. B. Allen are doing very well. They swarmed last week and were hived without other than ordinary labor. We may soon expect to have fresh honey added to our list of local products and bills of fare."[79] In an ironic twist of fate, then, Allen reversed the westward migration of bees by taking bees east. Thirty years later, the *Arizona Daily Star* reported that the bees were still doing well, "both domestic and those swarming in the mountains, . . . by which the entire territory and part of New Mexico and Sonora were seeded."[80]

By 1879, significant progress had been made by farmers addressing soil erosion. Clover, once thought to be a useless weed, became a lifeline for many beekeepers. In Kentucky, E. E. Barton introduced sweet clover into the Pendleton County region. Soil erosion had been a serious problem in Kentucky since the Civil War, and Barton wanted to introduce clover to help combat the problem. Of the twenty-five or more species of clover *(melilotus),* Pellett only mentions two biennial clovers—one with white blossoms and one with yellow—and he was not specific about which one Barton used.

The Anti-Bee Monopolist

With all these advances in bee inventions and social conventions, it was bound to happen that there would be a cultural backlash from the old honey hunters. Pennsylvania, long a haven for bees, was one of the first states to organize beekeeping associations. The invention of the Langstroth hive changed the ancient art of honey hunting, however, and even though the majority of beekeepers applauded Langstroth's discovery, a few die-hard honey hunters were ready to register their complaints.

In a letter written to the *Pittsburgh Leader,* an anti-bee monopolist wrote the following: "I can remember the day when any good, enterprising man, with half an eye, could always have half a dozen bee courses laid out, and at least a tree or two of honey for barter to a neighbor whenever he needed a bushel of potatoes or a side of bacon for his family. In those days, a 'bee course' was legal tender, almost."[81] The anti-bee monopolist continued his tirade by explaining how the barter system worked: "Why, just think of it! I have traded a bee course for a Bible, for a load of pumpkins, for a colt, for a couple of shoats and even for store goods, taking an order as good as the cash in payment."[82]

The anti-bee monopolist also talked about the services that were bartered for bee courses:

> I have seen the day when, if I wanted a couple of days' plowing done, I could go to any man of family in the township and tell him I had a good bee course, could give him the lines sure, and get my plowing done as handsome as though I owned all the cattle in the country. Everything's changed. . . . Everybody wants to speculate and monopolize things nowadays, so that a quiet man can't get along without pretty hard rubbing sometimes. You can't even take out a hound or two for a little harmless sport any more without being hauled up for trespass for somehow or another somebody's bound to inform on you. But of course, bee hunting suffers like everything else that was originally intended for man's benefit.

The *Pittsburgh Leader* published the old bee-hunter's letter but reserved the right to disagree with his antimonopolist stance, claiming instead to think that conventions are good for beekeepers, "but as the bee hunter treats the subject rather originally, he is permitted to have his say."[83]

Old World Customs in an Industrial Culture

Even as beekeeping was becoming increasingly more commercialized, more newspapers were printing accounts of wild swarms and tanging, which suggested a lingering curiosity and respect for the insect, even though the country was becoming more secular and more industrialized. In Alabama, for instance, the Somerville *Falcon* reported a rather "curious circumstance." It was rather warm and sunshiny and a young lady sat in the parlor playing the piano, with all the windows thrown open, when a swarm of bees, attracted by the music, entered a window and settled on the piano.[84]

In New York, a state commonly associated with the city, a longer article, titled "Hiving Honey Bees: The Remarkable Conduct of Some Farmers' Boys and Girls Explained," contains two good honey-hunting stories. The article begins with a description of a "bare-headed woman, with hair streaming in the wind, thumping a tin pan" and followed by young men and women, clanging on pots and pans. The city folk were startled. The group did not cry out or shout, but kept looking up, regardless of the brambles, briers, or marshy places.

The article contains an explanation for tanging: The tin-pan party was found grouped about a tree. A sturdy youngster was sawing off a branch on which the bees had settled in a big brown cluster. As he sawed, a young man explained, "When they're swarming . . . they will light on the first tree they meet with in their flight as soon as they hear the rattle and clatter of the tin pans."[85]

The writer continues: as the group saunters home with bees in tow, another story gets told about an Ontario & Western railroad man staying at Squire Knox's house. While in the middle of dinner, "one of the help, a deaf and dumb man, made a sign to the Squire, and grabbing a tin pan, ran out of the house. The Squire grabbed up the tin cover of a kettle, and then everyone grabbed up all the tin utensils they could find, and rushed after the Squire, each one beating on his tin."[86] The railroad man was convinced his host and hostesses were crazy, especially since Squire Knox, who came home empty-handed, didn't explain the process of tanging.

In this newspaper article, the New York writer emphasized a classic conflict between agrarian and city folk. The rural folk ignored gender-defined boundaries without even realizing it or feeling the need to communicate such an old European tradition. As such, they risked being labeled crazy by outsiders.

They were not bothered. Their love of bees was such that they stopped at nothing to try to coax the bees home. Furthermore, the railroad man, associated with progress and civilization, didn't look very sophisticated because he was also an outsider, so separated from the environment that he could not appreciate the bees.

There is no shortage of narratives about bees attending church. There are very good stories handed down from medieval times in Russia, England, Wales, Germany, and Ireland about the bees attacking sinful parishioners. However, one of the funniest incidents in America occurred on September 23, 1885, in Maryland. On that particular Sunday, the Wesley Chapel had some unexpected guests.

> The congregation of Wesley Chapel had tied up their teams, and settling into their seats on Sunday morning, when they found out that swarms of bees had taken possession of the church. They made the discovery so suddenly that they didn't have a

> chance to escape before the bees began business with indefatigable industry and persistence.
>
> It is estimated that the proportion of stings to the second in the fifty seconds that it took the [words unclear] to make the distance from the front pew to the door was seven to one. Some of the slower movers failed to cover the space in less than a minute and a half.
>
> The heroic pastor waited long enough to announce that the church was untenable, and that no services would be held that day. He was the last one to escape, and bears a proportionate number of honorable wounds.[87]

These nineteenth-century newspaper articles demonstrate a couple of important customs being defined in America: women worked with bees on a daily basis; people documented the ever-increasing conflicts between civilization and nature; and newspapers recorded a reporter's healthy respect for the agrarian lifestyle and the humor that can be produced when it clashed with the ever-encroaching industrial one.

Tall Tales

Wallace Stegner once said that before the Civil War, the West was not a goal; it was a barrier. Not content to confine their dreams, Americans relied upon "brute force" to impose their agricultural ideals upon the arid landscape.[88]

Americans had an unfortunate tendency to brag about these exploits in a genre known as the tall tale. The West cultivated tall tales. This genre was a peculiarly American invention, according to none other than Mark Twain. It is distinctive in two respects as far as story-telling traditions go: "the story must be told gravely about impossible circumstances, and it must finish with a well-timed pause, if it is to be successful."[89] As bees swarmed and settlers invaded the West, their encounters were bound to produce tall tales, the likes of which this country has not appreciated since.

Nowhere were tall tales as prolific as they were in Texas. Bee swarms and railroads show a classic formula of agrarian versus industrial conflict. In 1882, a young man remembered hearing tales from a section foreman who had been "helping build the Texas and Pacific Railroad west . . . when construction reached the Brazos River. They built the bridge and then began working up a gorge, laying the track as they went. While blasting out the rocks in this gorge, they uncovered a bee cave."[90] The exposure to the sun causes the wax to melt. "It flowed like lava between the rails and over them." In fact, as the story goes, workers had to sand the rails so the trains could get through.

A whisky-drinking, curiosity-seeking entomologist, Gustaf Wilhelm Belfrage, who sent insects to the chief museums of the world, single-handedly made the Round Rock region in Texas "bee-cave mad."[91] Having found a ravine filled with millions of bees, he started a cedar brush fire. "For two days, the fire was kept burning, and bees continued to pour out with the smoke."

Predictably enough, on the third day, no bees came out. This change in fortune has been an ominous sign in stories for centuries. In this story, however, the end was ominous for the bee hunters, for the fire that they had set burned the wax and honey instead. In fact, "the sinkhole turned into a belching furnace. For four days and nights the fire raged. The heat calcined the whole hill. . . . Today, the place resembles a long-deserted lime kiln."[92]

As late as 1894, Gen. John R. Baylor relayed a bear hunter's tale about a bee cave that "was worth a store full of coats . . . room after room was almost filled with long white curtains of the purest brush-blossom honey, some of them fifty feet high from top to bottom."[93]

But Jim Jones was the ultimate Texas honey hunter during the late nineteenth century. When launching an attack on a local bee cave, he hauled a pair of blacksmith bellows and rigged it with two hundred feet of hose to blow out the sulfur smoke. He employed another beekeeper to do the extracting, and both tied a honey extractor to a mesquite sled. Three hundred pounds of extracted honey—pure *huajillo* honey—was the net amount of one day's work.[94]

All good things must come to an end, and just as assuredly, Jim Jones had to die a bee hunter's death. The Southern Pacific Railroad crossed the Pecos River, and the bridge that crossed it was the longest and highest in the West at that time. About a hundred feet beneath the railroad ties were hanging combs that were too snugly attached to the wooden ties to be jarred loose. There was nothing to do but build a rope ladder underneath the bridge to get to the honey. This Jones did, but his mistake was leaving his hungry, honey-eating hound on the railroad bridge. Because the weather was exceedingly hot, honey flowed at the slightest jarring movement—except, according to Dobie, "instead of dripping down, it dripped *up* the rope ladder. This can be understood only by remembering that the rope went *up* from the honey instead of down."[95] Thus, Jim's hungry hound commenced to licking the honey as it dripped up the rope, only to lick so much he gnawed the rope bridge in two. Jim Jones in quick fashion dropped to the rocky floor of the Pecos Canyon, dying a bee hunter's death.

By the end of the nineteenth century, there were few recorded incidents about Indians and honey, suggesting that for them, the progress of America into the Plains was anything but funny. However, a first-generation American named Herman Lehmann, whose father immigrated to Texas with Prince Solms in 1842, wrote a captivity narrative about growing up with the Apaches. Herman was only eleven years old when the Lehmann farm was attacked by Mescelero Apache Indians in the 1870s. Herman was adopted by the Apache chief and became a warrior, taking great pride in attacking the Comanche Indians and buffalo soldiers. He also took pride in providing food for his fellow Apache brethren. When gathering honey on a raid, he was lowered into a bee cave: "This was a cloudy, damp day and the bees were out of humor, but I went naked in among them, filled my bag with nice new honey and comb, and hooted to be drawn up. This performance was repeated several times, and in this way we secured enough honey to last for a long time. There was plenty of honey left in the cave for the bees, too. Far back in the cave could be seen great clusters of comb honey, much of it black with age, and the storage must have been going on for years."[96]

Although Lehmann's narrative suggests a benign relationship with the bees, Colorado honey hunters take the exact opposite approach. In Gridley, Colorado, the *Herald* described "Bee Rock," located a half mile northwest of South Butte. The rock is about fifty by a hundred feet wide at the base and is from one to six feet wide on top. Nonetheless, a party of men determined to rob the hive, and the resulting newspaper article, "Storming a Bee Castle," is a hilarious read of men using mining equipment in order to defeat the honey bees. Military language is used metaphorically (perhaps even satirically) throughout the article. "The men were supplied with powder, fuse, drills, bars, with which to assail the stronghold. Few of the invaders had nerve enough to cross the bridge but three of them got over and fired a blast. The result was a cloud of bees that made them retreat. . . . Next day, the assault was renewed, and after a lively battle of three hours, the bees were defeated. The dead bees filled three grain sacks to overflowing . . . the party found a solid mass of honey in the comb two and a half feet thick."[97] Words such as *assault, retreat, battle,* and *defeated* suggest that this particular bee hunt was done aggressively. The dead were even counted, and to the victors went the spoils of beeswax and honey.

In a less place-specific article, writer Roger Welsch recorded a tall tale told by a pioneer Plainsman, which, judging from the content, may have been told during the late 1870s or early 1880s. The teller brags about taking a herd of bees from Nebraska to Texas on the hoof: "The next morning I made a noise like sage brush in full bloom and as the bees came out of the hive, the consignee counted them and found that I hadn't lost a single bee. We branded 'em and notched their ears that very afternoon."[98]

1880 to 1900

Women were still making headway into the beekeeping community. Hunter MacCulloch published the first beekeeping book written for women in 1884: *How I Made Money at Home, with the Incubator, Bees, Silkworms, Canaries, Chickens, and One Cow; by John's Wife.* Its inscription read: "Any woman will, if she can—any woman can, if she will." John's wife is introduced in this how-to manual. She provides an overview of bees, equipment, hives, honey, and wax production. It's not exactly clear who John's wife is, but she, as a character, does have a voice in the text: "For bee-culture is woman's work. It brings her more than money: the open-air exercise that the care of bees calls for, will in nearly every case cure most of her bodily and mental ills."[99]

Even the *Ladies Home Journal* featured an article advising women to consider the "agreeable, healthful, and lucrative employment" to be found in beekeeping.[100] Although the article acknowledges that men tend to be beekeepers, the writer suggests that "women ought to be better beekeepers than men, for they have, usually, a gentler, finer touch than men." The article continues with a list of other qualities such as patience, confidence, and absence of fear. Interestingly enough, the article ends with a pragmatic note: "Though millions of pounds of honey are produced every year, yet honey is practically unknown to the great body of people. There are abandoned farms north, east, south, and west, and there are tons of honey on these farms running to waste; and at the same time, there are thousands of women: pinched by want, wearied by toil, who could earn, with the help of the bees, more than they earn now."[101] This writer addresses the plight of many women widowed or left by men to explore the West. The idea presented in the article is that beekeeping can take the place of the domestic ideal so widely promoted by magazines (including *Ladies Home Journal*) during the postwar years. Interestingly, the encouragement comes from within the female, not male, community.

For instance, in 1896, Jenny Atchley encouraged people to begin beekeeping in Texas: "If you are going to make a beekeeper you must study your honey resources. . . . There is less competition in the bee business than in many other lines, and we ought to have our free, open unoccupied fields stocked with bees, that some of the many thousand tons of honey going to waste may be saved."[102]

During the 1880s, Lucinda Harrison was a large-scale commercial beekeeper in Illinois and published details of her bee dress.[103] No less beneficial to women were settlement schools started in the 1880s to teach people subsistence arts. Pine Mountain and Hindman were both important for providing an education to rural mountain people, but also for preserving folk arts, such as keeping bee gums or tracking bees.

The most poetic writer about bees was Emily Dickinson. Although her reserved stance, isolation, and reclusiveness pervade her other poems, I find the bee poems social, flirtatious, and very feminine: "Because a Bee may blameless hum / For Thee a Bee do I become / List even unto Me."[104]

Dickinson's bee poems emphasize metamorphosis and a realization made after the possibility for freedom has passed. The epiphanies in these bee poems are more subtle than her other poems. For instance, in "Safe in their alabaster chambers," the meek members of the Resurrection miss out on the "Light [that] laughs in the Breeze, / In her castle above them, / Babbles the bee in a stolid ear . . . / Ah! What sagacity perished here."[105] The "sting" in this poem is that the people miss out on the wisdom that comes from living outside the safe walls. Another poem begins with the line, "Could I but ride indefinite / As doth the Meadow Bee." Like the previous poem, the writer shifts focus to bodies enclosed in dungeons: "So Captives deem / Who tight in Dungeons are."[106]

Dickinson's bees are free whether they live in a hive, in a cell, or on a prairie. Even though the American prairies were thriving before bees arrived, Dickinson's poem about their relationship is a good one:

> To make a prairie it takes a clover and one bee
> One clover, and a bee,
> And revery.
> The revery alone will do,
> If bees are few.[107]

More than one Native American would have disagreed with this poem's content, for bees surely meant white settlers were imminent. But Dickinson is not to be faulted for never touching a Langstroth hive. She represents a consciousness of bees in the American feminine mind-set. She understood their work ethic and how laborious it was to produce something as sweet as a tiny drop of honey. Her bee poems are exactly that: drops of wisdom at a time often characterized by sentimentality and melodrama.

Children also benefit from the emerging children's literature market. Frank Stockton, the editor of the popular children's magazine *St. Nick,* wrote *The Bee-Man of Orn* in 1887. This book perpetuated the notion of a bee hunter being old, ugly, and cranky, but of paramount importance to the industrial American society. In his journey to discover his identity, the Bee Man is not fooled by society's conventions or temptations. Yet when a young baby is threatened by a dragon, the Bee Man uses his bees as a weapon. In this children's tale, Stockton completely reverses the typical American hero: young, technologically advanced (with some type of weapon), and by the end of the tale, richer for his adventures. The Bee-Man of Orn ends up poor and remains a beekeeper, but a very happy one, as is his society around him.[108]

For many Victorian American children, the nursery rhyme "Sing a song of sixpence," while charming and memorable, did not match their existence. There were no kings or pies filled with blackbirds. But for many, the image of the queen "in the parlour / Eating bread and honey" was their first introduction to the delights of the liquid. She is a perfect counterpoint to the king, who was "in the counting house / Counting out his money." Another proverb often featured in children's books of this time taught them the importance of timing:

> A swarm of bees in May
> Is Worth a load of hay;
> A swarm of bees in June
> Is worth a silver spoon;
> A swarm of bees in July
> Is not worth a fly.

Children were expected to understand the agrarian economy in which such a saying would have had meaning. In *Alphabet of Country Scenes* (1885), the rhyme book is notable for an illustration of the Langstroth hive instead of the more common bee skep in earlier children's books.[109]

Old World customs and superstitions about bees lingered in the public consciousness for at least two reasons: European immigration to America had not slowed, and newspapers recorded and thus perpetuated European superstitions about honey bees. As discussed earlier, two prevalent customs—that is, telling the bees of the death of the bee master and dressing hives in mourning cloth to prevent the bees from leaving—remained very popular in the nineteenth century. According to the article in an 1890 issue of the *Courier-Journal,* some people would even invite the bees to the funeral. The *Courier-Journal* also reminded its readers that bees must not be sold. "To sell them for money is considered a most unlucky proceeding," the reporter warns, "but they may be bartered away and all will go right."[110] The article suggests trading a swarm for a pig or a bushel of corn. But anything less, and the bees' "self-respect is touched, and they refuse to work for an owner who has bought them."[111]

More ominous superstition was that bees foretold death in a family. Swarming on a piece of dead wood was considered a sign of impending death. The reporter offers the story of a wife who died in childbirth. The husband "accepted the blow philosophically because he said they had been warned of the event a fortnight before the confinement. The woman went to the garden and saw that their bees, in their act of swarming, had made choice of a dead hedge-stake for their settling place."[112] The reporter added a footnote to the story: "It is more than probable that the prediction brought about its own fulfillment."

Even though American popular culture was mythologizing the bee hunter, tanging parties, and old European superstitions, significant inroads were being made in scientific beekeeping. During the 1880s, Kentucky beekeeper J. S. Reese developed the industry's first bee escape, an aid in harvesting, and enjoyed high yields of honey by using migratory practices. According to Kentucky bee historian William Eaton, "They operated as many as 425 colonies for several years until foulbrood hit the area, wiping out the life's earnings of several beekeepers almost overnight."[113] Foulbrood was (and remains) a serious threat to beekeepers.

Many states continued to form beekeeping associations. The Kentucky State Beekeeper's Association started in 1880, Iowa's started in 1885, and North Carolina's started in 1887. One of the first recorded instances of a young beekeeper was at the North American Beekeepers Society in Lexington, Kentucky, in 1881. Mary Nottnagle of Lexington, a seven-year-old, gave a recitation on honey bees, which was "warmly applauded."[114]

Mormons lamented that their state legislatures wouldn't know a "honey bee from a yellow jacket" when it came to funding money for an inspector.[115] The beekeepers lobbied and won out. By 1890, they had one—for the entire state of Utah.[116]

Many lively debates also took place during this time. Many beekeepers, who used the bee magazines and conventions to discuss patent infringement laws, wrote to express support for Langstroth, who at that time was still engaged in a lawsuit over his hive design. From Selma, Texas, a progressive beekeeper named L. Stechelhausen wrote in 1888 that he was very much in favor of Langstroth, but the German partisan support for Dzierzon should be noted, for German beekeepers were very loyal to their country's scientists. It took work on the part of German Americans such as Stechelhausen to convince newly arriving German beekeepers to adopt the Langstroth hive:

> Within the last few years I have written articles for different German papers, in which I explained the advantages of our American hives and management; this caused Dzierzon to advise us not to use a hive with frames to be manipulated from above, because by taking out the first frame, the bees would be rolled and killed; and other beekeepers had other objections. Nevertheless, our hive system is gaining more friends in Germany. No one can honor Dr. Dzierzon more than I do, but to call him the inventor of a frame hive is just amusing, when he used every occasion to speak and write against frames.[117]

The controversy concerning hives had important ramifications because many people tended to want to buy cheaper, smaller hives. The result was that often too much honey was taken from hives for the bees to survive the winter in the northern states. Or the hives would become crowded, causing the bees to swarm.

A Kentuckian named George W. Demaree announced his famous swarm control plan in 1892. Demaree realized that the reason for swarming was overcrowding. When the queen detects such pressure in the hive, she quits laying eggs and prepares to leave the nest. This transfers energy away from honey production. Demaree proposed to transfer the combs containing brood from the brood chamber to an upper story with a queen excluder. One comb containing brood and eggs was left with the queen, and the balance was filled with empty combs to provide ample space for egg-laying.[118]

Because this method required so much time if beekeepers had small hives, many transferred to the larger hives, which cut down on the swarming problem. "It was a start back to sanity in beekeeping," notes Pellett. It would take many years and many colony losses for people to realize the "drastic" effects of Demaree's swarm control plan, but it was an important reaction to the small hives so popular at the time.[119]

Blue waters wash up to Oahu from every vantage point. Bird of paradise plants pose nonchalantly in the most mundane places: gas stations, hotel courtyards, sidewalks, and city parks. Hawaii blooms with plants and people. At first glance, then, the *kiawe* (or mesquite) tree is the ugly duckling in Paradise: it is a short, squat, scrappy little plant eking out an existence in the drier parts of the island. But this plant is responsible for a minor but profitable beekeeping industry in Hawaii.

Hawaii was annexed as a territory in 1894, and its system of buying "bee rights" was unique. Because most of the land in Hawaii was divided into large tracts, beekeepers would have to approach the managers of a tract, buy the rights to agreed-upon locations, and arrange for payment. Although often the payment was in honey, at times beekeepers paid a fixed sum and assumed the risks associated with drought, disease, or other reasons for which there might not be a crop. If another beekeeper placed his colony within the limits of the prescribed area, according to "A Brief Survey," "It can end in only one way, since the beekeeper who has a right there has the advantage. . . . The moral right of priority claim, which so many beekeepers advocate, has small place in the manipulations of territory in Hawaii, where the beekeeping companies pay for what they get and insist on getting it."[120] Although Hawaii didn't officially become a state until 1959, it exported honey to the mainland as early as 1894. Beeswax was exported in 1897. These modest successes were indicative of how well the complex society—German, Hawaiian, American, French, and Japanese—worked together in Hawaii.

In the absence of a common political party, religion, or language, many beekeepers found a niche in the emerging American industrial landscape through their love of bees. Beekeeping helped reaffirm common bands of friendship, love of landscape, practical farming, and similar cultural practices, such as wedding celebrations and funeral arrangements. By the end of the nineteenth century, American beekeepers formed a solid community, even though major ethnic, geographical, gender, and racial differences remained. When the nineteenth century ended, Americans had restructured beekeeping so that it could be profitable and manageable. Four inventions transformed beekeeping from a cottage industry into a major commercial industry: the Langstroth hive, the smoker, the extractor, and the comb foundation maker. All four have stayed essentially the same. The commercial beekeeping industry developed simultaneously with large-scale agriculture immediately after the Civil War. With trade routes reestablished within the country and without, American farmers started to develop regional crop patterns that would define the twentieth-century migratory commercial beekeepers.

For all the emphasis on industrialization, American communities were determined to carve out religious or racially diverse utopias. For these people, beekeeping was as much about goodness of life and religious freedom as it was about profits and bank ledgers. Their songs about bees remain some of the best testaments of their faiths and of this country's efforts to secure the right to life, liberty, and the pursuit of happiness in spite of the high price of human lives the Civil War exacted.

The independent honey hunter coexisted with the settled, profitable beekeeper. Mass publications perpetuated the image of honey hunter as an ideal just as society around him was changing to a corporate culture. The honey hunter remained an important but shadowy figure as the advertising, pop culture market, movie, radio, and children's literature industries developed in full force during the twentieth century. However, the economic circumstances that supported him would no longer be in place.

John Burroughs, a noted bee hunter, best reflected the transition period between the nineteenth and twentieth centuries. In "An Idyll of the Honey-Bee," he issued an invitation to go bee hunting: "If you would know the delights of bee-hunting, and how many sweets such a trip yields beside honey, come with me some bright, warm, late September or early October day."[121] But he also best described the changing appetite of the nation: "It is too rank and pungent for the modern taste; it soon cloys upon the palate. It demands the appetite of youth, and the strong, robust digestion of people who live in the open air."[122] By the twentieth century, the openness of the open air was very much in question, and honey bees and their keepers were at the center of the controversies.

Part Four

Requeening a Global Hive

Chapter 5

Early Twentieth Century
Industrialization, 1901–1949

The only consistent thing about bees is their inconsistency.

—*Dr. C. C. Miller*

For much of the early twentieth century, America was balanced between a pastoral ideal of sustainable agriculture and an emerging commitment to a new form of agriculture that would characterize twentieth-century America. The two ancient symbols of sustenance—bees and cattle—were in forty-eight states as well as in the territories of Alaska and Hawaii. A third symbol—the train—made migratory beekeeping, a uniquely American twist to the agricultural industry, possible. The nomadic trade route patterns that were established continued to be in place and provided a framework for transforming America into an industrial countryside, to borrow Steven Stoll's phrase. In *Fruits of Natural Advantage: Making the Industrial Countryside in California* (1998), Stoll suggests that five factors merged in the twentieth century to create a highly industrialized agricultural landscape: capital, science, innovative farmers and orchard growers, an independent yet inextricable relationship between farmers and government, and a solid hierarchy between owners and laborers that generally divided along class and racial lines.[1] All of these factors played into a general agricultural trend for farmers to specialize in certain crops. Yet at the bottom of this industrial countryside was the honey bee, whose ability to pollinate would not be fully appreciated until after World War II, but who nevertheless made the agricultural transformation possible in the nineteenth century and visibly successful in the early twentieth.

Furthermore, advertisements promoted this concept of industrial agriculture: those who worked the American soil would be rewarded. Advertisements with honey bees promoted pastoral images and children, suggesting innocence. No one would starve in the modern Canaan; advertisements said so. Yet it was hardly an ideal country. Industrialization affected the bee industry just as it did other businesses. Commercial pollination, World Wars I and II, the queen rearing industry, a price support system, and imported bees improved American beekeeping. Making money in honey became the norm; beekeeping was no longer a sideline interest. However, beekeepers reeled in the wake of diseases, chemical pesticides, and zoning restrictions. As with many twentieth-century industries, science and technology could outpace an average beekeeper's ability to apply new knowledge before changes were made again.

Honey hunting did not disappear in modern America. Economic depressions, wars, and health reform movements encouraged people to find value in such activities as finding bee trees, making candles, and cooking with honey. The bee hunter, although marginalized, was still an admirable person in many poverty-stricken regions, and thus, a small literature of bee hunting tales still existed in these areas, promoting independence and sustainability.

The honey bee remained an important image in the twentieth-century arts. In addition to reflecting more traditional values of industry and thrift, modern artists conveyed the complexities of racial and gender roles in American society. The honey bee—with its contradictory abilities to create honey but deliver a powerful sting—was a perfect metaphor by which to explore in movies, music, and literature the complexities of romantic relationships or traumatic experiences. In fact, artists embraced the honey bee for its cathartic potential.

The honey bee was "the step child of agriculture" throughout the twentieth century, according to *American Bee Journal* editor G. H. Cale.[2]

Nevertheless, in an increasingly technical and secular world, many Americans still harbored respect for bees, although bees and their keepers were decreasing in numbers.

Progressive Beekeeping

Progressivism marched through America with a vengeance, and beekeeping did not escape its influence. Although there had been health reform movements in the United States, the first law to protect consumers was passed in 1906, and much credit should go to honey producers such as Charles Dadant. Until this bill was passed, unscrupulous producers could sell honey that had been adulterated with sugar without fear of legal recrimination. Honey adulteration occurred often enough that the market suffered in the States and abroad. The problem was aggravated by the plethora of honey varieties produced in such a vast country. Hawaiian producers, for example, often provided two types of products: mesquite and honeydew. Mesquite honey, considered a delicacy, brought (and still brings) high prices. But honeydew is an inferior product, often made from sap or other sugary substances that foraging bees happen to collect. Pre–World War II bakeries liked to use honeydew because it has great baking properties. If the two liquids were mixed, however, mesquite honey would deteriorate in quality.

With an extensive network of alliances in both France and America, Charles Dadant had organized political pressure to protect beekeepers and consumers as early as 1878. Although initially indifferent to the question of honey adulteration, A. I. Root gradually came to his friend's support.[3] In fact, one summer, "Three of the National Beekeeper's Association Board raided a prominent Chicago operation producing adulterated honey. Eugene Secor, [Herman] Moore, and [George] York were commended in the Chicago papers and all the bee press for their actions," reports Kim Flottum, current editor of *Bee Culture*.[4]

After years of petitions and organization, beekeepers generated support among the states and territories for the bill to pass in Congress in 1906.[5] The unity was widespread and consistent. The Pure Food Act was in part due to the organized presence of beekeeping associations that formed after the Civil War to educate people and share resources. But these associations were redefining their roles. According to Frank

Pellett, the emphasis of beekeepers had changed from education to marketing in the twentieth century: "By 1910 the idea of doing big things in the way of advertising and selling honey began to be felt. . . . The following year, at Minneapolis, the society definitely launched on the plan of a business organization instead of an educational one, as it had always been. It was the beginning of the end of the golden age of beekeeping."[6] In 1911, the Round Rock Beekeeping Association in Texas accurately summed up the attitude binding beekeepers: "Progressive means doing something worthwhile. There is no standing still in this or any other business."[7]

The pollination industry developed alongside the transportation industry. In 1912, the Texas Department of Agriculture bulletin provided pictures of hives on tops of city buildings, a city window unit, and a series of photographs of migratory pollinators using horse-and-buggy outfits, Ford Model Ts, and railroad cars. Beekeepers traveled to farms throughout the year, thus systematically earning money and providing a much-needed service to commercial farmers.

Nephi Miller was especially successful at migratory beekeeping. In 1894, Miller traded five bags of leftover oats to a neighbor for a colony of bees. By 1904, he had enough confidence in his beekeeping skills to quit his job on a wheat-threshing crew and become a full-time beekeeper. In 1906, to help out with the labor, he hired a twelve-year-old Swiss immigrant named A. H. Meyer, who would go on to become a commercial beekeeper in South Dakota.[8] He was doing rather well for himself until he came up against a problem.

"Beeswax—and what to do with it. There seemed to be no way in which all the impurities could be removed from it," writes Nephi's biographer, Rita Miller. Miller explains, "Industry at large was discovering the value of byproducts that previously had been thought useless and were allowed to go to waste. It was true of the steel, meat packing, lumber, and mining industries, for example."[9]

Beeswax has always been useful, but an efficient method of cleaning the wax of impurities was difficult to come by. So after borrowing some money from a local bank, Nephi Miller decided to go to California in December to learn how to render beeswax in clean, smooth cakes. While there, he was fortunate to meet a commercial beekeeper named M. H. Mendelson, who taught Miller the steps of cleaning beeswax. Mendelson would also hire women to do queen rearing, finding them to be better workers than men at such a delicate task. Miller never forgot Mendelson's kind instruction and often followed his mentor's examples with his own employees. But more importantly, Miller could see that his bees, stationed in the cold Wasatch Mountains in Utah, would thrive in the warmer California winters. But how? Railroads were his answer, just as they had been for Migratory Graham in the nineteenth century.

After knocking on a lot of doors, Miller finally reached an agreement with Union Pacific Railroad to transport bees. In 1907, Nephi Miller moved his bees south to Colton, California, which had plenty of orange groves. Then he moved them north to Utah during the summer in order to get two seasons out of them. Because "train men were horrified at the thought of hauling bees," Miller would leave the railroad cars open, and even rode along with his bees.[10] He is still considered the father of commercial beekeeping in Utah. Many beekeepers followed his example until automobiles and interstates made cross-country pollination easier and more affordable.

Cars were especially important in the emerging pollination industry. According to one editorial written in the *American Bee Journal* in 1903, "There is no question as to [cars'] advantage in one respect—they will never get frightened, run away and break things by being attacked by cross bees."[11] Horses waiting to transport bee colonies from the railroad to farms would be stung by the bees and become very difficult to handle. Although railroads could move large quantities of bees, they could not guarantee comfort of the bees. If railroad cars were left in the sun, the bees would literally cook if neither ventilation nor water were quickly provided. According to one beekeeper, Howard Graff of Snohomish, Washington, "If it was a warm day or the railcars were delayed, many hives would melt down, and honey would run onto the floor of the boxcar."[12]

Roger Morse cites the first occurrence of bee rental for pollination in New Jersey during 1909.[13] Interestingly enough, the pollination industry offered a double-edged sword, according to Morse: "Development of the automobile helped create both a problem and the solution: New suburbs cleared out woods and other natural habitats where bees had built their hives, but at the same time, the improvement of roads and spread of car ownership made it easy for domesticated bees to be taken wherever they were needed."[14]

Women were welcomed to the industry with open arms. Jennie Atchley began publishing *The Southland Queen* (1901) in Texas; she explained bee diseases, discussed new equipment, and provided information to beekeepers throughout the vast state of Texas.[15] Its *art nouveau* motif, with a woman bedecked with roses and laurels, emphasized its queen bee editor. Curving trellises, elongated fonts, and an attractive woman created an exotic look for the publication, especially as far as beekeeping journals go. Even though the magazine did not last long, Atchley extended a tradition started by Mrs. Ellen Tupper in Iowa during the 1880s.

The two bee business giants—A. I. Root and Charles Dadant—offered unflagging support for women beekeepers and factory workers. Both men employed women in their comb foundation and candle factories and published many letters and articles written by women beekeepers. In fact, perhaps A. I. Root's riskiest stance as an editor was to publish an article about Mrs. Booker T. Washington's all-girl beekeeping club in Tuskegee, Alabama.[16] Margaret Murray Washington's beekeeping club included women so confident with the bees that they weren't wearing bee veils in the pictures. "[A. I. Root] caught some flack from some 'less enlightened' readers, and lost some subscribers in the South," writes Kim Flottum. "But he kept many from the North, so it all balanced out, or so he thought."[17]

A. I. Root also published *Beekeeping for Women,* written by Anna Comstock in 1905. Comstock, an entomologist, wrote the book—which still holds value—for beginners.[18] When ordering from catalogs, Comstock advised that the "catalogs are so delightful that, if the purse is long enough, [one] feels inclined to order a specimen of everything listed. However, this is not the best way to do."[19] Easier said than done.

Still, maternal concerns were a real factor in a woman's ability to be a beekeeper. In "Beekeeping for Women," written by Mrs. H. D. Woods for the 1911 Texas Beekeeping Association, she stressed that beekeeping is a great occupation because it doesn't interfere with being a mother: "Texas is the promised land, while to many it is a howling wilderness just because they do not prepare to gather up the milk and honey while both are being wasted. . . . But, Brother Bee-Keeper, beekeeping with me is a side line, for the work that interests me most is the boys and girls, the raising of noble sons and daughters according to God's laws."[20]

5.1. Margaret Murray Washington's Beekeeping Ladies, from Booker T. Washington's *Tuskegee and Its People: Their Ideals and Achievements* (1905). The wife of Booker T. Washington taught beekeeping classes in addition to cooking, gardening, and other vocational skills at Tuskegee. The bee classes were more popular with women than men. Most women in this picture are not wearing bee veils.

By 1917, Frank Pellett, editor of the *American Bee Journal,* stated that "so many women have taken up beekeeping that there is no longer any novelty attached to a woman beekeeper."[21] In fact, Pellett featured an article about three former schoolteachers—R. B. Pettit, Mathilda Candler, and Emma Wilson—because they became successful honey producers, not hobbyists. Each left the classroom, that traditional bastion for professional women, for the freedoms and demands of beekeeping. Pettit, Candler, and Wilson had no regrets about leaving the classroom, although they approached the industry in different ways. R. B. Pettit plunged in immediately, buying one hundred hives, but then again, her brother, Morley Pettit, was the provincial apiarist of Ontario. She merely extended the tradition.

Mathilda Candler drifted into beekeeping slowly. Her plans to study art in France were disappointed when a family member invested her savings unwisely. Although bees had been a sideline hobby, she gradually increased them so that she was able to make a profit. When bees provided enough money to pay the bills, Candler abandoned the classroom.

Emma Wilson "forgot her resolution to return to the classroom at

5.2. Beehive Beverage advertisement, courtesy of Utah State Folklife Archives, Utah Arts Council. By straddling the ad, the young man creates a triangle that automatically focuses attention on the advertisement.

the earliest possible moment" when she went to visit her sister and brother-in-law, the famous beekeeper C. C. Miller. In fact, Wilson was as much the star of the Miller household as C. C. Miller: "Miss Wilson is a strong-minded individual and no mere echo of Dr. Miller." Wilson enjoyed a national reputation as a good beekeeper; she wrote for the *American Bee Journal* for many years.

Women were good crew workers in the field. M. H. Mendelson, the same man who taught N. E. Miller to render beeswax, hired women to be queen rearers for his large California 1,400- to 2,000-colony apiaries. Women "were careful and painstaking," and because no heavy lifting was involved, it was a win-win situation.[22] His queen-rearing crews wore overalls and heavy jackets—there was not a dress-wearing woman in the bunch.

Bees and beauty go hand-in-hand in advertisement campaigns during the early part of the twentieth century. Twentieth-century advertising changed nineteenth-century perceptions about women and cosmetics. Although honey was a common beauty aid, Victorian culture associated cosmetics with prostitutes. Feminine beauty should come from within, it was thought.

5.3. Beehive Beverage advertisement, courtesy of Utah State Folklife Archives, Utah Arts Council. This soft drink used honey in its recipe. Ads emphasized energy, health, and mobility.

Beeswax and honey were fundamental ingredients in many cosmetics. And models often used products such as lipstick and creams in advertisements. The Mormon company Beehive Beverages promoted its soft drink with the slogan "Good and Good for You" as early as 1914. A beehive flanked by pretty women appeared in advertisements. Later, when Beehive Beverages used photography in its ads, the women wore short skirts, bobbed hair, and makeup. The hive, long known as a symbol of industry, is in front of an automobile, a classic American symbol of mobility. Two women on either side of the automobile promise beauty, vitality, and security—provided one bought the beverage!

Reflecting a new democratic principle, twentieth-century cosmetic surgeons and beauty industries suggested that even if one had been born unattractive, one could become attractive because cosmetics such as lipstick were affordable and available.[23] When women decided to wear makeup, especially after World War I, honey and beeswax became prized commodities.[24] Beeswax is still used in cosmetics to make lipstick, rouge, and cold creams, but beeswax was especially used during the early twentieth century. By the 1940s, American industries would use 4 million pounds of imported beeswax because American beekeepers could not produce enough to meet the demands of consumers.[25]

By the time America entered World War I, beekeepers were in the midst of a major paradigm shift. The four major beekeeping inventions from the nineteenth century—the Langstroth hive, the Quinby smoker, the comb foundation maker, and the extractor—had fundamentally changed a cottage industry to a commercial one, but the ways in which these shifts would affect commercial beekeeping were still being streamlined. According to F. L. Aten at the 1911 Texas Beekeeper's Association Progressive, "Bee culture has made a great many strides. . . . We have greater profits from the bees, more money per hive; less expense in handling them; neater package; better grading; advertising honey as a health food (which it is); holding off the market when it is glutted."[26]

World War I

When World War I erupted in Europe, honey prices increased because sugar was more expensive and more difficult to obtain. Honey became the default sweetener, as often had been the case during wars. Physician Bodok Beck estimated that prewar prices for extracted honey cost five cents a pound, but during World War I, "honey sold in carlots from twenty to twenty-five cents a pound."[27] An article in a 1916 *Bee Culture* discusses the importance of honey in the European trenches: "The war lords in Europe, when it came to the matter of rations, soon discovered that honey, an energy producer, was much cheaper than sugar (also an energy producer), and consequently honey has been going into the trenches, and is going there still. Apparently, only medium grades are being used, because they furnish as much energy per pound as the finer and better-flavored table honeys that cost as much or more than sugar."[28] The world shortage of sugar had catastrophic effects for the bees in Germany and Austria. During the winter of 1918–1919, many of their bees starved while their keepers were serving in the trenches. According to G. H. Cale, one reason for the high prices of honey was that at the beginning of World War I, "we were exporting in the neighborhood of five million to ten million pounds of honey to Germany and other warring nations."[29] Furthermore, Russia was a big importer of beeswax for church purposes. When the war affected that market, the beeswax market dropped initially, only to increase in 1917 upon the United States' entry to the war.

World War I was the first modern war, and it differed from previous methods of warfare in at least two ways: the increase in facial wounds (the result of the nature of trench warfare), and the increase in victims of shell shock. The severity of these injuries was not just a physical problem but a social one as well. The British and American governments genuinely feared that veterans would not want to resume their careers when the war ended. Morale was a serious consideration, and both governments promoted cosmetic surgery as a way to help veterans recuperate their sense of confidence in addition to basic communication skills. Even though surgeons would make facial functions their first priority, according to scholar Beth Haiken, they were equally concerned about their patients' appearance when they began to operate.[30] In fact, the British government sponsored American cosmetic doctors' travel and board while in England.

But both governments shared a larger concern for those patients who would not recuperate quickly from shell shock or those veterans whose injuries were too severe to be easily masked by cosmetic surgery.[31] Beekeeping was considered a viable alternative career because a veteran could work alone, at a slower pace, and still contribute to society. Honey prices were also high, so the beekeeper could earn a good salary. In both countries, programs were established to help disabled veterans adapt to their injuries by teaching them beekeeping.

In America, a group of seven extension workers was hired to teach better beekeeping methods—George Demuth, Dr. E. F. Phillips, Frank Pellett, Jay Smith, E. R. Root, and M. I. Mendelson.[32] Walter Quick wrote a pamphlet in 1919 titled "Bee Keeping to the Disabled Soldiers, Sailors, and Marines to Aid Them in Choosing a Vocation."[33] Both the *American Bee Journal* and *Bee Culture* published articles that encouraged veterans to learn beekeeping. F. Eric Millen wrote a feature article in the *American Bee Journal* about Harvey Nicholls, an Iowa beekeeper who had lost both of his legs in a boiler explosion. The article assured veterans, especially those who had been disabled, that "you can [be a beekeeper] if you will."[34]

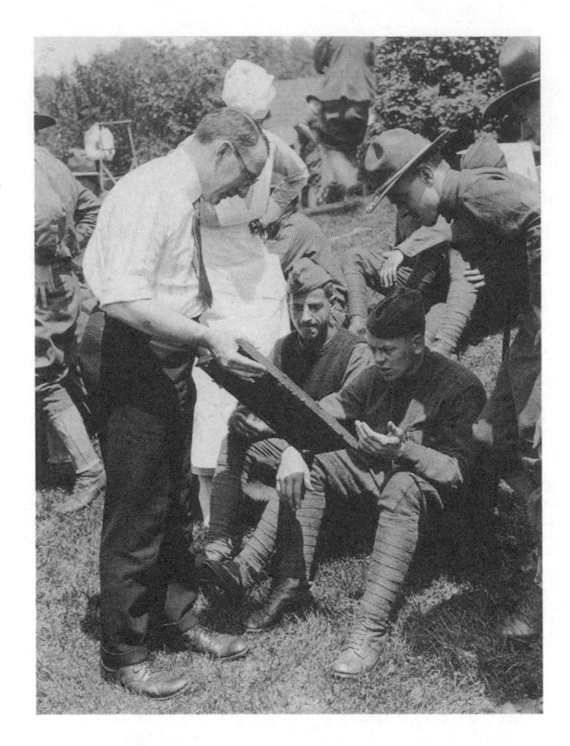

5.4. E. F. Phillips examines a frame with World War I veterans (1919). Courtesy of *Bee Culture*. As part of the government's effort to train vets to become beekeepers, Phillips explains the complexities of the bees to soldiers.

E. F. Phillips was the most ardent advocate of beekeeping programs for veterans. Although he readily admitted that he was reluctant to teach beekeeping to the average person because the failure rate was so high, he was excited about teaching veterans returning from France. Phillips emphasized to his veteran students that "the most important part of a beekeeper is above the neck." Indeed, in the pictures, one veteran had an amputated leg, one had a jaw shot off, and some were suffering from shell shock. Some of these veterans ran to the woods when the hives were opened, although Phillips does not explain, except to say that they had been through gas attacks in the war.

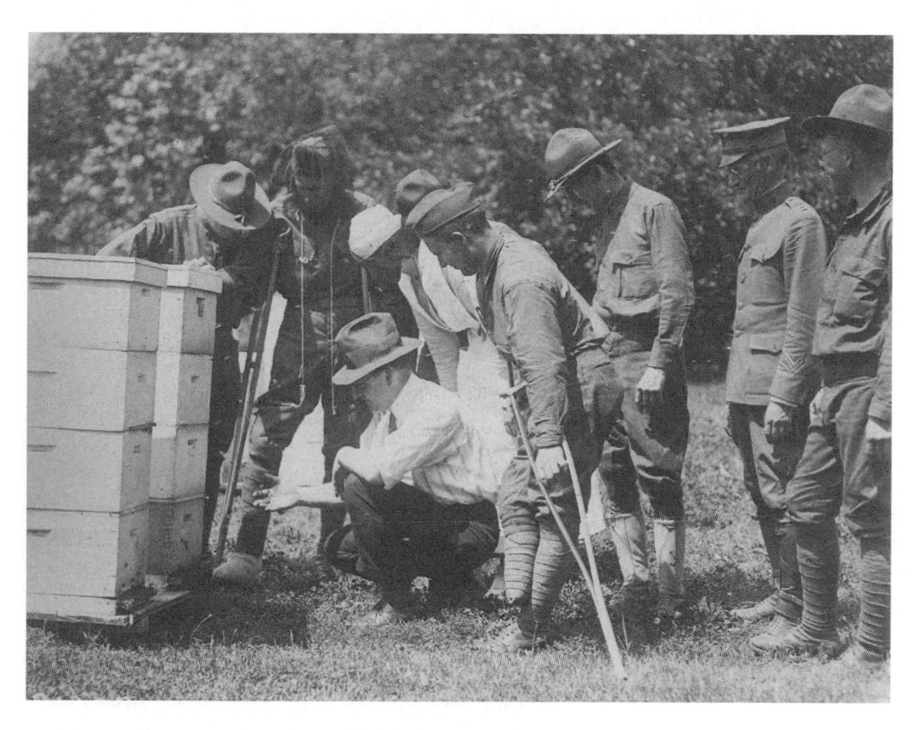

5.5. E. F. Phillips teaching hive basics to World War I veterans. Courtesy of *Bee Culture*. Coaxing a bee to land on his hand, Phillips is surrounded by vets recuperating from war injuries. Their serious looks have changed to smiles. Most don't wear veils.

On the whole, Phillips was very positive about the U.S. government's efforts to train veterans to become beekeepers. He felt that good commercial beekeepers were needed, and if returning veterans were not able to continue their previous careers, "I, for one, shall be glad to see them take up the work."[35] Still, according to F. Eric Millen, "We cannot expect many of our disabled soldiers to take up beekeeping. It requires special tasks and aptitudes. No man who dislikes the work can succeed in keeping bees."[36]

The *American Bee Journal* was especially prominent during World War I because of the role Camille Pierre Dadant played in helping the Belgian and French beekeeping communities rebuild from the war damage. Because of his French connections (he was ten when his father immigrated to Illinois), C. P. Dadant coordinated relief efforts for beekeepers whose supplies and bees were destroyed by the war, sugar shortage, or harsh winters. In 1919, he reminded Americans, "We are deriving profit from high prices and those high prices are due in great part to the suffering of Europe."[37] Dadant also published a letter from Mr. Crepieux-Jamin, the former associate editor of *Revue Internationale d'Apiculture*, which read: "The disaster defies description. . . . Wherever things have been left standing after the bombardment, the Germans burned them or blew them up, cutting down the fruit trees and shrubbery. It is a desert. . . . Some beekeepers whom I know would be glad to begin over with a colony or two. The busy hum of the bees would undoubtedly encourage them. What they hope for is to be helped with a fresh start, the first few colonies."[38]

Donors, their letters, and dollar amounts were published in the *American Bee Journal*. In the 1920 volume, Dadant reprinted one letter from Noah Bordner: "Enclosed is 1.25, for which please send *ABJ* for one year to some beekeeper in France or Belgium, with instructions to pass it on from one beekeeper to another for I think they need good literature as much as supplies."[39] Dadant responded enthusiastically and offered to substitute a French edition of Lorenzo Langstroth's book if French/Belgian beekeepers couldn't read English!

Dadant also published pictures of European devastation. These pictures were just as powerful as the letters. The *American Bee Journal* featured a little town called Grand Pre, France, using before and after photos to highlight the damage done to towns and bee skeps. So involved was he that he personally traveled to Europe to help the Belgium beekeepers. For Dadant's efforts, King Albert of Belgium awarded him Knight of the Order of the Crown.

Gene Stratton-Porter's *The Keeper of the Bees* (1925) is a young-adult novel about a World War I veteran. The novel's protagonist, James Lewis McFarlane, returns to California, wounded in body and spirit. He suffers from the same malaise afflicting a generation whose Victorian ideals about women, community, and religion were destroyed by modern warfare: "I began with a gnawing fire in my breast and a bitter blackness in my heart and brain." The Bee Master and a little girl called Scout teach Jamie beekeeping, until the Bee Master goes into the hospital. The beekeeping chores then fall on Jamie's shoulders. The discipline needed to work with bees helps Jamie recover physically and spiritually. When approaching the Bee Master in his hospital room, Jamie describes the peace he feels while working with bees: "If [I] could earn money like that, if [I] had a garden of wonder to work in, if I could earn the Bee Master's confidence, if I could daily make worthwhile friends, if I had a wife, if there were going to be a child to bear my name, what was the use in dying? There might be something very well worthwhile I could do in the world." Before the Bee Master dies, he leaves his garden and bees to Jamie. By the end of the season, Jamie confronts black German bees with confidence and thus exorcises the ghosts of the German soldiers. His recuperation is twofold then: "He was so nearly a well man that he was beginning to use his left arm without realizing that he was using it. . . . Every day was a day of work that he loved in a location that he loved."[40] Finally, McFarland opens his heart to women and God again. But those steps take a long time and require the discipline and beauty of beekeeping.

Although Jamie McFarland learned beekeeping *after* the war, many men left their colonies to be soldiers—H. F. Carrilton of Illinois, for example. During the two years Carrilton served in the army, his bees were neglected at home. Fortunately, he was released in the spring, so he had time to clean up his colonies before the nectar flow.[41] Similarly, William W. Mitchell and Irvin Powers owned bees before serving in World War I. When both men returned from the European front, they became honey producers in Idaho.[42] When the war ended in 1918, a young man named Charlie Heckman decided to go to California. While there, he met Nephi Miller, who "worked right along with us on the job," Heckman wrote later. Their partnership was a good one: Heckman worked in the California, Utah, and Idaho branches.[43] And serving in the Army Signal Corps was a young man from Michigan named Walter Kelley, who would commercialize the bee supply industry in the South, following in the steps of his predecessors A. I. Root and Charles Dadant.

Arguably, the most revered war veteran was Reverend Francis Jager, who served as an interpreter for the Red Cross in the Balkans. Later he served as a chaplain in the Serbian army and then the United States relief commission. It was in that capacity that he met King Alexander I of Yugoslavia, who awarded Jager the Order of Sava in recognition of his service to the Serbian people.[44] When Jager returned to the States, he took charge of the Division of Bee Culture at the University of Minnesota. He became a prominent spokesperson and teacher throughout the world, in part because he was fluent in so many languages.

Commercial Honey Production

Sweet clover (*melilotus*) had an indirect effect on honey production during this time, for this plant transformed the cattle and bee industries. Like the honey bee itself, sweet clover was imported into the United States, although how and when are not known.[45] During the nineteenth century, many farmers and beekeepers initially furiously debated the use of clover; some states even passed laws to prosecute those farmers planting the "noxious weed."[46] But by the twentieth century, this debate had ended with the overwhelming successes associated with the plant, including those of Iowa farmer Frank Coverdale, who was able to reap a double windfall with fat cows and honey.

Farther south, E. E. Barton introduced sweet clover, which transfers atmospheric nitrogen to the soil, to help replenish the soil in Kentucky. The poor soil had caused many tobacco farmers to go bankrupt, and when their farms were foreclosed, lawyer-beekeeper Barton would buy them and plant sweet clover. Sweet clover seed production became an agricultural staple in Falmouth, Kentucky. Pendleton County became known as the "home of sweet clover." From 1916 to 1917, Pendleton County beekeepers produced 500,000 pounds of honey and sold queens.[47]

But these farms were nothing compared to North Dakota, Texas, Washington, and California farms. Large citrus, apple, and almond orchards, to say nothing of clover, watermelon, cucumber, alfalfa, and squash crops, needed or were aided by pollination. In the Red River Valley, North Dakota, the sweet clover bonanza was one of the state's first major industries. In 1922, there were "40,000 acres of sweet clover in one North Dakota county, Grand Forks."[48] The Red River Valley merges into the Missouri River Valley, which is legendary cattle range country. In addition to sweet clover, the bottomlands of the Missouri were excellent for growing alfalfa hay for the Herefords that populated the ranges. George Greig of Garrison, North Dakota, was a pioneer bee man of the Missouri Valley, and he set a good example for a long line of North Dakota beekeepers.

Social Reforms

While farms and agriculture benefited from progressive land policies, cities adopted progressive social reforms aimed at children and teenagers. Institutions such as schools and community centers promoted beekeeping as a disciplinary measure. The YMCA taught beekeeping courses as early as 1917.[49] The juvenile court system also used beekeeping as a social reform. When a "shiftless" and "incorrigible" Boston youth appeared before the Children's Court, he would be sent to "the Farm." He was the despair of the authorities until he was sent to bring a colony of bees back to the Farm. "Mac" was stung so many times he decided that he would have to learn about beekeeping. Of course, Mac fell in love with the bees, and he left the Massachusetts corrective facility a well-respected individual.[50]

Sex education was also important during this period. According to Julian Carter, "Between 1910 and 1940, sex educators faced the problem of how to teach young people about sex without encouraging licentiousness—in other words, how to educate them in a way that would guarantee premarital chastity and marital monogamy."[51] Thus, for many educators, images of birds and bees were appropriate for young audiences because pollination was less overtly sexual than many animals they would have seen. Although pollination is the movement of pollen between plant parts, bees act as the mediator of the pollen as they collect nectar. And birds tend to be nesting creatures. Thus, both birds and bees were used to promote "safe" sex without anybody having to address potentially embarrassing topics. According to Carter, the good intentions of the progressive reformers were also hampered by ambivalence: "For fear of encouraging licentiousness by giving people too much information about sex, educators frequently sacrificed the 'scientific truth' in favor of whatever kind of information or teaching strategy was most likely to encourage premarital chastity and marital monogamy."[52]

Photographers were important participants in the progressive period, for often they recorded exploitation in pictures that could be used as powerfully as written speeches or letters. The famous child labor pho-

5.6. Vermont farmer John Spargo checking the hives with his son. Courtesy of the Library of Congress, photo by Lewis Hine (1914). A sociologist by training, Lewis Hine used photography to teach Americans nontraditional education methods, such as the outdoor classroom.

tographer Lewis Hine was able to get a few pictures of Vermont beekeeper John Spargo with his twelve-year-old son in a 1914 series titled "Work that Educates." Hine traveled around the United States in order to channel his social concerns through photography. In this series, Spargo's son weeds the garden, tends the chickens, and watches over his father as Spargo checks the bees. After having worked for the National Child Labor Committee for ten years, Hines surely must have enjoyed taking these pictures in which a child was not working long, dangerous hours in a factory.

Speaking of social issues, beekeepers were not immune to the racial problems affecting the South specifically and America as a whole. On a trip through Mississippi and Alabama, Frank C. Pellett did not mince words when describing the racial differences in the South: "In most of the southeastern States, the white people live in the towns, while the colored population tills the soil."[53] Although he was greatly impressed with the vast unused fields of white and sweet clover, Pellett still cautioned the northern beekeeper about relocating to the South: "The northern man who goes south, expecting to show those who have lived there for years a better way of doing things, is not likely to succeed very far. The southern people have problems different from ours, and they know better how to approach local conditions than a stranger does."[54] One such problem was the low wages for labor-intensive work. When he visited in 1917, Pellett writes, "The price of cotton was so low it hardly paid expenses."[55]

Two years later, when the price of cotton increased because of World War I, Pellett would revise his opinion of blacks in Mississippi. The tenant farmers improved their standard of living so that they were no longer a "dejected looking lot."[56] Pellett mentioned black beekeepers at the Greenville beekeepers' meeting. When he drove out to Richard Grant's apiary, Pellett was impressed at how neat it was and that black women were involved in bottling the honey. In the photo caption, he states that there were some very good beekeepers among the "colored population."[57]

Perhaps because beekeepers and psychologists rarely cross paths, Pellett did not mention Charles Henry Turner, a black schoolteacher who published over seventy articles about honey bee research.[58] Turner, who was one of the first scientists to prove that bees can perceive color and patterns, was recognized by well-known scientists for his numerous achievements. In 1910, while conducting research during his summers, he theorized that bees made "memory pictures." In 1911, he published a second paper that extended his work on color vision. Although his peers in France accorded him great respect, Turner never achieved recognition in the United States. Nor did Karl von Frisch cite Turner, even though, according to Charles Abramson, "Professor Turner was probably better known in Europe than in the United States."[59] Abramson cites major scholars in animal behavior who admired Turner's work: Margaret Washburn, E. L. Thorndike, T. C. Schneirla, and Eugene-Louis Bouvier. So von Frisch's silence regarding his colleague is, as Abramson concludes, "perplexing."[60]

Nonetheless, African Americans were making inroads in the South via bees. When Booker T. Washington opened the Tuskegee Institute in

Alabama, beekeeping was part of the curriculum. A. I. Root donated materials and bees to Tuskegee. When Washington died, Root wrote a fine tribute to the man and his mission in *Bee Culture*.[61] And at least in Washington, D.C., blacks had begun to congregate in a place called "The Bee." The building was featured as part of an exhibit titled "The American Negro" for the 1900 Paris Exposition.

Both *Bee Culture* and the *American Bee Journal* provided good articles about Asian American beekeepers. In one *Bee Culture* article, feature writer Louis Wahl chanced upon a laundry owned by a Chinese American in Colorado. Leung Chung had bees in the walls of his laundry, so Wahl helped him transfer the bees to a hive.

When Leung Chung invited Wahl to his ranch, the two men developed an easy rapport. The conversation turned to Chinese laundries. Using his own experience as an example, Leung Chung explained the need for better labor laws in the United States: "I come here, I hire for cook for 50.00 a month. I work one month all right, so the next month, boss say, I can't pay this month; you work a month more; then I pay you both, or else you go. See? Then I work another month, an no mon'; then I find out he got none. . . . I then start wash. First week I make 7 or 8 dol; then if some don't pay, I don't lose all."[62] In the midst of World War I and industrialization, beekeepers (and bees) flourished in the first part of the twentieth century, and beekeepers themselves showed more progressive attitudes in both the bee yards and toward women and minorities.

The Roaring Twenties

When World War I ended, there were a number of side effects that affected American beekeepers. Most immediately, honey prices dropped because sugar was again readily available. Women, the primary grocery shoppers, tended to buy sugar instead of honey. Beekeepers felt betrayed at the sudden reversal in the nation's tastes. In fact, Dr. Bodog Beck groused, "The modern housewife uses 'honey' only . . . as a word, when she is anxious to have a new fur coat, an automobile, or jewelry."[63] Consequently, a few producers lost the financial incentive to maintain their hives. Foulbrood became a primary concern for state politicians trying to protect the domestic honey industry.

Furthermore, the interwar years were marked by isolationism as America withdrew from global politics and markets. Beekeepers were affected by this political move. Before the war, beekeepers had enjoyed great freedom in importing different varieties of bees: German black (although these imports were few), Egyptian, Cyprian, Syrian, Carniolan, and the already popular Italian bee.

But importation of foreign bees came to an abrupt end when the 1922 Honey Bee Act was passed by Congress with little opposition.[64] This act was considered a proactive measure to stop the spread of the Isle of Wight disease, which had begun in Britain. The Atlantic Ocean would be a barrier. The 1922 Honey Bee Act was fueled by fear of diseases, immigrants, and contamination.[65]

Another federal law requiring all hives be inspected and outlawing bee gums followed on July 1, 1923. Bee inspectors were authorized to have hives that showed signs of American foulbrood (AFB) or European foulbrood burned; beekeepers with bee gums had to buy modern hives. The law stated, to quote bee inspector Eugene Shoemaker, "It shall be unlawful for any person to keep or maintain honey bees in any hive other than a modern, movable-frame hive which permits thorough examination of every comb to determine the presence of bee disease."[66]

These political actions meant that the United States bees would not be exposed to foreign bee diseases or pests. The Honey Bee Restriction Act also eliminated keepers from importing other strains of honey bees.

The Queen-rearing Industry

An indirect result of World War I was the queen-rearing industry. Given the world shortages in sugar and honey, beekeepers needed to be able to rear queens in order to increase their colonies. Improved transportation meant bees could survive shipping. Pellett states: "The business of queen breeding developed slowly, but steadily, until the time of the great expansion that came with the World War and the spread of sweet clover over the Plains Region. Since that time the growth has kept pace with the development of the package business, which has expanded so rapidly. Much of the present day prosperity of the queen breeder is due to the popularity of sweet clover as a farm crop."[67] Sweet clover had been planted in the large states such as North Dakota, and beekeepers could count on good honey crops if they placed their hives in these fields. As a practice, it is advantageous for the beekeeper to use commercially produced queens with known traits for several reasons: it can help breed resistance and it can help a beekeeper produce profitable colony dividers.

Queen rearing is a relatively simple process. In any hive, a queen will lay many eggs, from which new queens can develop. While these queens-to-be are in larval stage, they are fed royal jelly, an extra nutritious food, and are tended to in special cells by nurse bees. This process allows for the queens' ovaries to fully develop.

In an isolated region such as Hawaii, beekeepers needed to be able to produce their own queens. Apiaries were divided among the cattle ranches. And when queens were quarantined as early as 1908, island beekeepers were forced to work with their existing supplies. Hawaiian beekeepers experimented with rearing queens that produced AFB-resistant offspring. Because AFB continued to plague mainland beekeepers, the hope was that Hawaii would be the best place to experiment with queens.

Because of the earlier seasons in the South, southern beekeepers quickly cornered the queen-rearing market because they could produce queens earlier as a result of favorable weather.[68] Warmer temperatures meant orchards and crops would bloom earlier, providing both pollen and nectar for bees. A young man named Harry Hyde Laidlaw Jr. grew up working with his beekeeper grandfather, Charles Quinn, in Texas. They had a method of queen insemination known as the Quinn-Laidlaw hand mating system. The drone would be held behind the queen until it was stimulated to act. However, many times the offspring were not quite what Laidlaw was hoping for. The hand system was also incredibly inefficient. The 1927 Mississippi flood acted as a catalyst that would promote Laidlaw's studies with bee insemination.

The 1927 Mississippi flood wasn't just any flood. According to historian John Barry, "From Cairo, Illinois, to the Gulf of Mexico and from New Orleans to Washington, D.C., all across the floodplain of the Mississippi River and beyond, the 1927 flood left a watermark."[69] When the waters had receded back within their banks, the remaining vestiges of King Cotton and the plantation South literally washed down the river—and took many beehives with it. In fact, a magazine published by E. G.

LeStourgeon called *The Beekeepers Item* broadcast the extent of the damage. In the June 1927 issue, Jes Dalton wrote the following: "Flood water may not be a beekeeping question direct, but in the marketing of package bees, it may be of more importance than at first thought. If a northern buyer sends his money south, and the apiarist has all his holdings swept away, or even holds his apiaries intact, but all his train service is shut off, or those which are shipped are detoured beyond the endurance of shipments, it becomes a problem of interest to at least those two persons." In the same article, Dalton said proudly that beekeepers would not "pass around any hat or sit on a corner with tin pans extended,"[70] yet he did request that people write their politicians to pass funding that would bring the entire levee system of the Mississippi River under the federal government's jurisdiction.

In the meantime, Louisiana state entomologist and apiary inspector W. E. Anderson made an appeal to the beekeepers in other states to send money, materials, and bees, specifically queens. In writing of the damage, Henry Stabe wrote to *The Beekeepers Item:* "Not only have they suffered losses of the crops, often all of their livestock, not to mention their bee hives and less valuable movables about their farms, but in many cases their homes; while in all cases, their fences, barns, bridges, automobiles, and in fact, everything that could not be loaded into a small lifeboat has been taken away from them. . . . Nine-tenths of the commercial beekeeping of Louisiana was done in this flooded territory."[71] Respond, beekeepers did. LeStourgeon provided a detailed list of accomplishments that only beekeepers could appreciate. Harry Wilson commandeered boats and barges to rescue hives and equipment before they were washed out to sea. The American Railway Express donated free transport of materials: "Here came box hives, log gums, bees in every style of hive or package that one had ever seen, hundreds and hundreds of them."[72] Furthermore, a "Great Camp just a few miles from New Orleans was established where bees, and queens and all those things needed to rebuild beekeeping was gathered together." LeStourgeon finished his article with an incredulous tone: "Beekeepers who six months ago saw their businesses swept away, no means of employment left, even their homes gone, today have their apiaries ready for spring."[73]

In short, if Steven Stoll's thesis about the network of factors between farmers and government can be reconsidered, the Great Flood of 1927 tested the industrial countryside and how it would respond to a natural disaster. Louisiana rose to the challenge in part, to summarize LeStourgeon, because there were four factors in place: Louisiana already had had a strong beekeepers association before the flood; Louisiana had a thriving entomology program at Tulane University and Louisiana State University; technological resources such as the American Railway Express and queen rearing speeded the process; and Louisiana had a labor force of people willing to help.

Even before the Great Flood, the United States Department of Agriculture had set in place plans to establish a laboratory specializing in genetic testing of honey bees in Louisiana with the full endorsement of the southern states. *The Beekeepers Item* published a doctrine from the Texas State Beekeepers Association requesting that "if such a laboratory is to be established in the Gulf Coast region, it be located in Louisiana for the purpose of studying the genetics of the honey bee, the selection and breeding of queens, and the problems of bee production."[74] The rationale was that the Gulf Coast bee breeders have "peculiar phases and complexities that have never been scientifically worked out."

Harry Laidlaw Jr. moved to Louisiana immediately to raise queens for the state. By 1929, he began trying to figure out why artificial insemination was so difficult for queen bees. While working on his master's degree, "Harry discovered the 'valve fold,' a tonguelike structure in the median oviduct of the queen that was preventing the proper placement of the semen-filled syringe during insemination," explained Robert Page Jr.[75] Once this crucial fold was discovered, queen rearers had more control of breeding.

A beneficiary of both the scientific advances and the people left without equipment was Walter Kelley, who had moved to Louisiana and opened a bee package and bee supply company in 1924. By the time Jes Dalton had visited him in 1928, Kelley had opened the bee supply factory. Cypress lumber made wonderful hives because cypress was less prone to water damage, and the warm weather benefited his bee package industry.

Although federal legislation and science was affecting beekeeping in complex ways, honey marketers promoted the product by using images of children and other races suggesting a simpler time. The Sioux Bee Indian princess is a prime example of a successful campaign. In 1921, the Sioux Bee Honey Association developed in Iowa when five men—Edward Brown, Clarence Kautz, M.G. Engle, Charles Engle, and Noah Williamson—banded together with $500 in capital and three thousand pounds of honey. The Indian princess was chosen as the company icon to reflect the company's location, Sioux Falls, Iowa. Many beekeepers have profited by belonging to this organization.

However, the Indian princess could be construed as a direct extension of nineteenth-century marketing techniques that, in the words of scholar Jeffrey Steele, "fixed the Indian in the past."[76] And to extend Jeffrey's Steele's argument, the tendency with such an advertising campaign is to suggest that all Indians were considered the same. Sioux Indians were not known for their beekeeping skills; in fact, they were nomadic raiders. The Sioux bee honey princess was an effective campaign, but one that was effective because it promoted Indian women in a glamorized fashion.[77]

The Cherokee, who maintained a strong agricultural history, were better situated to become beekeepers. In his 1920s study, anthropologist James Mooney had this to say about the Cherokee: "Bees are kept by many of the Cherokee, in addition to the wild bees which are hunted in the woods. Although they are said to have come originally from whites, the Cherokee have no tradition of a time when they did not know them; there seems, however, to be no folklore connected with them."[78] Similar to the Sioux Bee Indian princess, a popular honey label printed by bee companies "fixed" the black child in the past, to borrow Steele's phrase. The ad featured a black child ready to eat honey on pancakes. The black child has one tooth missing, and the image reflects a time period when stereotypes about rural blacks were accepted.

Black people were eating honey, but they were not limited to the stereotype featured in the honey industry ads. The Black Renaissance flowered in the 1920s in New York, attracting many southern black people who wanted better living conditions and higher wages. When they moved North, many blacks found comfort in the Black Muslim movement, which was beginning to establish strong urban communities. Black Muslims, using the writings of the prophet Mohammad as their guide, followed strict dietary laws, including using honey rather than sugar whenever possible.[79]

Honey labels also featured or appealed to children. In one advertise-

5.7. Honey advertisement featuring black child, courtesy of Dadant Company. Children were especially popular images on honey labels because they suggested innocence and drew attention away from the stinging characteristics of honey bees. This honey label was standard at many bee supply companies and reflected many stereotypes of the early 1920s. Contemporary ads are more neutral and sensitive to race differences.

ment from the 1930s, Hopalong Cassidy, the hero in a series of stories by Clarence Mulford, promoted spun honey, the letters being spelled by his lasso. Other advertisements featured idyllic scenes, suggesting that honey could indeed be one's link to an earthly paradise. By focusing on the land or children, advertisers hoped that shoppers would be more likely to forget about their fears of bees. In fact, after he was well established in the bee business, Walter Kelley decided to use his own image to promote his products, figuring that people would buy products from a face they could identify with. His head was superimposed on a bee, which

5.8. Honey advertisement featuring Hopalong Cassidy. Courtesy of *Bee Culture*. This character was wildly popular with children because it perpetuated the formulas of the Westerns. In a clever marketing idea, the lasso creates the letters, roping the buyer in for a sweet deal.

would make people laugh. According to Mary Kay Franklin, "His remark was 'Make a fool of yourself and people will remember you.'"[80] Much later, in reflecting on the importance of advertising, Kelley remembered back to his first business, Gulf Coast Apiaries: "As a beginner that sounded big and important to me but I soon found out that few people understood the meaning of the word and some asked if we kept apes."[81]

Writers of children's literature were an important force in how bees were seen in American culture. Although much of early twentieth-century children's literature was "considered a sugar coating around the bitter pill of education," according to Eva-Maria Metcalf, it was an important and influential market.[82] In Philip Mason's bibliography, he provides a thorough list of children's publications about bees.[83] Children's literature also appeared in the *American Bee Journal* and *Bee Culture.* In fact, women wrote many of the children's sections in these magazines. These sections were not really concerned with rearing future generations of beekeepers as much as rearing civic-minded individuals.[84] During the 1920s, mass publication opportunities arose, as did discretionary income for white parents. Children's literature continued to be a solid—though marginalized—market in each decade after the 1920s, although the market fell just after the 1929 stock market crash.[85] The most notable writers of children's literature about bees are Maurice Maeterlinck, Margaret Warner Morley, and Mary Geisler Phillips.

Although the children's literature and honey markets plummeted after the stock market crash, the beeswax candle market flourished. Before the 1929 crash, the A. I. Root Company expanded its focus from manufacturing beekeeping supplies to candles. The current CEO of the A. I. Root Company, John Root, told the following story about his great-uncle Huber: "Huber and the local Catholic priest loved to play tennis. While they were out on the new courts in Medina, the priest started complaining about a new shipment of candles. He suspected that the candles were not made of beeswax. He asked Huber if the company would want to make some candles for the local church."[86] Although the church had been buying beeswax candles, the local priest wasn't satisfied with the quality.

In a summary in *The Catholic Encyclopedia* (1907), Dr. Bodog Beck explained the importance of beeswax to the Catholic Church: "No Mass could be sung without pure beeswax candles. Wax, extracted by the bees from flowers, symbolized the pure flesh of Christ which He had received from his Virgin Mother; the wick signified the soul of Christ and the flame, His Divinity."[87] The Paschal (Passover) candle at Easter time has been associated with the Resurrection of Christ since at least A.D. 384. St. Augustine, who was influenced by the patron saint of bees St. Ambrose, wrote in *De civitate dei,* "Among bees there is neither male nor female. . . . The wax of the candle produced by the virgin bee from the flowers of the earth is as a symbol of the Redeemer born of a Virgin Mother."[88] Suffice to say that beeswax matters in the Catholic Church. So when the Medina, Ohio, priest wasn't happy, Huber Root decided that the company should redefine its role in the beeswax business "in spite of the fact that it was launched during the Great Depression."[89]

5.9. Women on the assembly line at A.I. Root candle factory. Courtesy of *Bee Culture*. Even before the Depression, women were employed as candle makers at Root and Dadant companies. Their hands were smaller and could finesse the wax easier. All women wore dresses and kept their hair up.

The Root company beeswax candles had several advantages over other candles. Their candles were hand-rolled before they were dipped into beeswax, and thus, they would rarely break. Because of this construction, if the candles arrived warped, church officials could remold them into shape. The Root Company also offered different sizes, which were priced for the poorer immigrant families from central Europe. Because churches had been sponsored by state taxes in their home countries, many immigrants were not accustomed to paying for candles. So Huber Root fashioned the tiny votive candles, priced at a penny, so poorer families could participate in the service and provide a small donation to their church.[90]

The candle market did not drop during the Depression, as did honey. "Beekeeping does better during hard times than other industries," according to the current editor of the *American Bee Journal*, Joe Graham. "People go back to church, and churches need candles."[91]

5.10. Women sewing bee suits at A.I. Root Bee Supply company. Courtesy of *Bee Culture*. Women were also employed as seamstresses, making bee veils and suits.

Having come in with a bang, the Roaring Twenties ended with a whimper—or seemingly so. But political developments occurred during this time period that would continue to affect America during the 1980s and 1990s: the federal laws meant that honey bees would not develop resistance to global diseases and beekeepers would have to keep in touch with the new industrial changes taking place in order to meet with federal inspections.

Honey bees were finally taken into Alaska, that last outpost of the American wilderness, in 1924. J. N. McCain brought the first hives into Alaska, although the bees did not make it through the winter. By 1927, McCain successfully overwintered seven colonies. He lived close to Anchorage and the Alaska Railroad. Although beekeeping didn't become an immediately successful enterprise, McCain's efforts were important. Honey could supplement the northern diets with a high-energy food.

Plus, Alaska enjoyed long summer days, flowers, and a hospitable summer climate. With this territory, the country's westward migration ended.

Overwintering was no longer a mystery to beekeepers by the time McCain had taken his bees to Alaska, although he had a difficult time keeping his bees alive. Packing the hives can be a necessary chore in the North, although it wasn't until the 1980s that the *American Bee Journal* reported that Evan Sugden's experiments with overwintering in Minnesota "found no significant difference between overwintering with light or heavy insulation."[92] However, wrapping is still necessary in most of Canada and the northen United States.

Overwintering success relies on two basic conditions: the bees are healthy and the hive has a strong queen. Bees also need plenty of honey to eat during the winter. According to Tibor and Daniel Szabo, "Weight losses during the winter are related to colony populations. Larger colonies use more food, were more populated in the spring, and produced more honey." Thus, Szabo and Szabo state: "It is essential to know the maximum possible weight loss of colonies during the winter months. Using the maximum possible weight loss one can estimate colony carbohydrate consumption and establish the optimal weight of the colonies prior to winter."[93] For those interested in commercial honey production, these factors make a difference in terms of profit and loss margins and costs of replacing colonies.

Finally, bees need access to the outside. They are exceptionally clean and will suffer from disease if they are not allowed to carry their refuse outside the hive. Diseases such as nosema and AFB can occur in the winter. One last word of advice about overwintering from Mary and Bill Weaver: "The precise temperature used in overwintering bees indoors is less important than keeping the interior of the bee storage area *completely* dark, so that the bees do not try to fly."[94] There are more factors to overwintering success, but it is no longer such a mystery.

The beekeepers' successes in the hive paralleled the country's successes and growing confidence in the early twentieth century. In fact, when downtown Chicago had to be rebuilt after the 1871 fire, city planners commissioned architects to create a unique style for the area, one that would mark the city as a symbol of progress and technology and also allude to its proud past. The Continental National Insurance Company commissioned Graham, Anderson, Probst, and White to create a building that would embody these values. The result: a skyscraper with a beehive supported by four buffalo. The beehive symbolized the industry of Chicago's people along with the many industries that had located to the heartland city. The building, erected in 1924, has become known as the Encyclopedia Britannica building because it was sold to the company soon after. Since its creation, the Britannica Beehive has been a reminder to its citizens about the value of hard work, seeing them through many difficult stretches in the twentieth century.

The Great Depression

After establishing himself as a humanitarian during the 1927 Mississippi flood, Herbert Hoover dismissed the Great Depression as a "crisis of confidence." But the crisis was more serious than that. The Great Depression involved unwise economic and agricultural policies, and beekeepers did not escape the debilitating effects that these had on honey bees and honey prices.

The Depression affected beekeepers in several ways. First, droughts occurred across the country. In Appalachia, the 1930 drought had more disastrous effects than the 1929 Wall Street crash.[95] Combined with the effects of strip mining and deforestation, the drought affected the bees' foraging materials. In Iowa, droughts caused many orchards to wither and die. In North Dakota, the dust bowl conditions were so bad that bees could not survive, much less make honey.[96] After so much careless land use and drought, many Midwest farmers changed from clover crops to better-paying grains.

Financial shortfalls took their toll on beekeepers. According to a 1933 *American Bee Journal* article, many Iowa orchard farmers were unable to pay for proper spraying, and thus, many orchards were chopped down. Just as the orchard growers neglected their trees, beekeepers neglected their hives. "Many boxes went without paint and repair. Some of the honeycombs were eaten up by moth and mice. Many colonies just plain starved and died," according to Woodrow Miller.[97] The federal farm policies did little to help small farmers, who did not have huge farms or much capital. Many beekeepers had to become a "self-reliant lot," who had to "muddle through" the Depression, in the words of one editorialist for the *American Bee Journal*. When he was in his eighties, Glenn

5.11. George Arnold exhibits a super of honey, Chaffee County, Colorado (1939). Courtesy of the Library of Congress, photo by Arthur Rothstein. Federal loans were given to farmers who could prove significant damage either because of neglect or poor growing conditions during the Depression. Note the optimism and pride Arnold displays in spite of his circumstances.

Gibson remembered the low prices that honey bought to a beekeeper: "In the thirties, good quality honey moved in carlots for under 3 cents a pound. Producer-packers delivered honey in 5 pound pails at 25 cents. Small side-line and hobby producers consigned honey to local grocery stores and received pay in groceries. Some producers dickered with hard-nosed chain store buyers. . . . In simple terms: It was a buyers' market."[98] In fact, Roger Morse remembered one Florida beekeeper who claimed to have dumped his honey in the Everglades so that he could at least use the containers again.[99]

Finally, the Farm Security Administration began to offer loans to beekeepers, but it wanted the agricultural damage documented. In Virginia, for instance, most of the honey bees had died from poison sprays, so pollination had to be provided by out-of-state bees. In 1939, photographers Arthur Rothstein and Russell Lee captured on film the difficulties beekeepers were facing. In Chaffee County, Colorado, George Arnold had suffered significant bee colony damage, all of which was documented by Arthur Rothstein in a series of photographs. Rothstein also went to Cache County, Utah, to photograph Donald Gill's beehives, most of which

5.12. Oakley Idaho Welfare Farm. Courtesy of Utah State Folklife Archives, Utah Arts Council. The beehive was used frequently during the 1930s to assure Latter-day Saints families that they would not starve in the midst of the Depression. Welfare Farms were set up in which people could work in exchange for food.

had been damaged.[100] These beekeepers were called "rehabilitation clients" and were approved for loans to pay for sugar, equipment, and bees. In San Antonio, Russell Lee captured a shot of a honey peddler, reading a newspaper, his cart stacked with jars. All of these photos were effective at convincing the Farm Security Administration that federal funds should be approved to help beekeepers, especially because by this time it seemed that America's involvement in World War II was imminent.

The Depression forced many people to reestablish social ties and policies. In Utah, the Mormon Church created welfare farms in order to provide a cooperative for families, according to Utah Folk Arts Council spokesperson Carol Edison.[101] The icon for the welfare farms, displayed on huge signs, was a big comforting beehive, which emphasized society, industry, and church.

At the N. E. Miller & Sons Honey Company, Nephi Miller, who was paying his workers $125 per month, had to call his workers together for the bad news. "He said the salaries would have to be cut in half; he would try and hold the business together and make it go. All the helpers stayed with him except one."[102]

The Civilian Conservation Corps was one of Franklin Delano Roosevelt's social programs that put young men to work on projects across America. The program would pay them a minimum wage, feed them, and train them in some type of engineering program. William Rowley ran away from home at the age of fifteen to work for the CCC to send money back to his parents. He credited the training he received through the CCC with helping him become an engineer, which he then used in World War II and with the Bureau of Reclamation Service. He became interested in bees after he returned from the war, fascinated by the orderly parallels between engineering and bees.

E. C. Honl was a hobbyist beekeeper before he joined the CCC. He went to bakery school, then used his hobby to help with his bakery. Finally, he became a full-time beekeeper after the war.[103]

One teenager, Clarence Dean in Myton, Utah, bartered honey for tuition at Westminster College. The college students ate the honey throughout the year.[104] On the opposite end of the social ladder from the college students were the "shellers," a group of people living on the Mississippi River. They raked up oyster shells (hence their nickname) to sell to button factories. The shellers depended on bee trees and log hives to supplement their diet, but bee inspectors would have to issue them citations if they did not hide their log hives in time.[105]

It is no surprise, then, that John Lockard's *Bee Hunting: A Book of Valuable Information for Bee Hunters* was so popular. Although it was originally printed in 1908, it was reprinted in 1932, 1936, and 1956. Lockard's preface clearly states that beekeeping was not his intention: "In the preparation of this work, it has been my aim to instruct the beginner in the art of bee hunting, rather than offer suggestions to those who have served an apprenticeship at the fascinating pastime."[106] Lockard appeals to people interested in lining bees, not buying equipment, land, or even working with a producer.

Although blacks were not highly visible beekeepers, the best twentieth-century writer to incorporate the honey bee image in American literature is Zora Neale Hurston. Hurston, a Florida native, moved to New York in the 1920s and worked closely with the noted anthropologist Franz Boas at Columbia University. In her fieldwork with turpentine workers in the Florida pine camps, Hurston learned valuable folk arts, songs, and aphorisms. Later, when she collected songs for the Works Progress Administration (WPA), she had occasion to learn how blacks used the honey bee in the blues. The result: *Their Eyes Were Watching God,* a novel in which the honey bee best describes the development of a young lady into a woman made powerful by love.

Hurston wrote this pastoral love story in 1935. The novel was based on her own tumultuous love affair with a man of West Indian parentage in 1931. Although she walked away from the relationship, Hurston admitted, "I tried to embalm all the tenderness of my passion for him in *Their Eyes Were Watching God.*"[107]

Having just entered puberty, the sixteen-year-old protagonist Janie grapples with her newly awakening sexuality: "While lying under a pear tree, she questions the complex roles of Nature taking place in the tree, the How? Why?" "Soaking in the alto chant of the visiting bees," she has a flash of insight. "The inaudible voice of it all came to her. She saw a dust bearing bee sink into the sanctum of a bloom; the thousand sister calyxes arch to meet the love embrace and the ecstatic shiver of the tree from root to tiniest branch creaming in every blossom and frothing with delight. So this was marriage!"[108]

Unfortunately, Janie does not experience these sensual qualities of marriage until she meets her third lover. Her first husband, Logan Killicks, is little more than a "vision . . . desecrating the pear tree." When Killicks decides to put a yoke on Janie and make her plow his field, she realizes the harsh inequities that can exist in a marriage. Janie promptly runs away with the man who would become her second husband, Joe Starks. Joe offers a new life, but dashes Janie's hope that he will become a "bee for her bloom." Joe is concerned about money and class status, not romance. Both of these husbands leave Janie wanting more from a relationship.

Only sweet-talking, long-legged Tea Cake Woods manages to tease out Janie's sexuality after years of a loveless marriage: "He could be a bee to a blossom—a pear tree blossom in the spring." Tea Cake, a younger man, entices the thirtysomething Janie to leave the all-black town of Eatonville and be his lover. When she returns to Eatonville, she is as regal as a queen.

Black writers found Hurston's book offensive because it did not do enough to promote social causes and fight racism. According to Richard Wright, Hurston's "characters eat and laugh and cry and work and kill; they swing like the pendulum eternally in that safe and narrow orbit America likes to see the Negro live: between laughter and tears."[109] However, this novel links black women with positive natural images in the American South. The power and sexuality that had always been associated with queen bees finds a creative outlet in a black female character.

Meanwhile, about the time that Hurston was finishing her book, Harry Laidlaw Jr. was busy unlocking an easy way to inseminate queen bees. While working on his doctoral dissertation, he made the process of artificial insemination easier by anesthetization of the queen. During his work on the queen's genitalia, he discovered the valve fold, a previously unknown part of the queen's anatomy, which acted as a "gate" to semen, and often, if the gate did not open, certain experiments with bee genetics would fail. Thus, Laidlaw redesigned the instruments used access the queen's sperm-atheca. Finally, the semen syringe was reconfigured to prevent backflow during the process.[110]

Then Otto Mackensen, whom honey bee geneticist John Harbo described as a "master with a metal lathe," improved the insemination devices to the point that some are still being used today. These devices would help inseminators bypass the valve fold, allowing semen to be injected into the oviduct. These practical developments in insemination during the 1930s and 1940s paved the way for the more complicated genetics work that would develop later in the century.

In the meantime, AFB continued to wreak havoc on bee colonies in the United States. It arrived on the Hawaiian Islands in the 1930s and decimated the declining industry on the Islands. The downhill trend in Hawaii had begun in the 1920s, when prices for honey dropped to pre–World War I prices. But mismanagement on the ranches left the bees in weakened conditions and thus susceptible to the disease. Fortunately, the two AFB-free islands—Lanai and Niihau—had been notified beforehand that AFB had been brought to the other islands and taken proper precautions. These islands remained safe from foulbrood.

A number of research institutions, most significantly Iowa Agriculture, had been devoting time and energy to finding a cure for AFB. O. W. Park had been committed to developing foulbrood-resistant bees until a sulfa drug was invented in the 1940s. Once this sulfa drug was patented, most experiments with disease-resistant bees subsided.[111]

Efforts to contain foulbrood were still being funded at state levels. The bee inspector generally followed the cattle inspector, which could be unpleasant for both farmer and inspector alike. Eugene Shoemaker was hired in the 1930s to inspect hives, to require that bees be kept in moveable hives, and to destroy any hives with AFB. Shoemaker put it mildly when he states that among any group of farmers, there were "those with hot tempers who spoke without too much thought."[112]

John Bruce was more specific about his adventures as an inspector: he got in fights with beekeepers, federal ration boards, and, of course, the bees themselves. Bruce said, describing one old man, "He held a double-barreled shotgun with both barrels cocked. He was so close that the barrels almost touched me as he held it aimed at my middle." Bruce was not an easily intimidated man. He also had to confront a German immigrant who had AFB in his hives with the following warning: "Mister, the state doesn't pay me to fight, nor does it condone it, but if you want to go out on the public road, off of your property, I'll take off my star and put it in the car."[113]

As if to counter the state inspector stories, Mary Louise Coleman offers a conventional feminine narrative in *Bees in the Garden and Honey in the Larder* (1939). When talking about the bee suit, Coleman cautions, "One would never take a prize in a beauty contest clad in that suit of heavy drill." Coleman reminds beekeepers that bees do not need the same amount of affection as other animals need: "It was a severe jolt to my ego that my precious bees never showed any recognition of my personality nor any acknowledgment of my being their keeper." In Coleman's view, bees helped her achieve the three Cs of cooking, cosmetics, and care of the house. In her book, the 1930s woman is one who has a foot in both camps of domesticity and agriculture. Coleman even extends hope that women may be able to find the cure to AFB: "Women have made their influence felt in beautifying the countryside throughout our land. May we all join forces to exterminate this American foulbrood."[114]

The radio provided relief through the dust bowls, droughts, bankruptcies, and poverty. The airwaves were free, and in the early days, radio stations offered time slots to a great diversity of music shows. Cultural historian George Douglas places the importance of the radio

5.13. San Antonio Honey Peddler, 1939. Courtesy of the Library of Congress, photo by Russell Lee. The full honey wagon, the availability of tin, the new hat—all these details suggest that World War II has not started yet. Plus, there is a nice contrast between the honey man, in his corduroy pants, boots, and leather jacket, and the elegant woman pictured in the newspaper.

in context: "The Depression was the making of the radio in more ways than one. Smaller incomes, recession, deflation, meant that more Americans had to watch their pocketbooks and spend more evenings at home before the fireplace. They would sit before the radio. . . . these little boxes became not only household utilities . . . they become a solace and comfort in the nation's darkest hour."[115] The radio was a powerful marketing tool, even more so than were the movies. Gospel music, which appealed to both black and white listeners, was an especially profitable product aired on radio. One of the most powerful cases in point is the Carter Family's *Honey in the Rock*, issued in 1937. This collection, which harkened back to the biblical promises of salvation in the midst of hard times, was so popular that it was reissued before the decade ended.

5.14. Busy Bee Home Social Club, Louisiana, 1940. Courtesy of the Library of Congress, photo by Marion Post. Nurse Marguerite White making plans for the next meeting with Rosie Patterson, assistant secretary of the Busy Bee Homeworkers Club on La Delta Project, Thomaston. A Red Cross pin pulls White's collar together.

Too, there's the sublime moment in George Gershwin's *Porgy and Bess*. "Here Comes De Honey Man" is a short vignette, but in this tragic work about love on the wrong side of town, the interlude acts as the "calm before the storm." The honey man was a conventional character in many towns and markets. In San Antonio, for example, Farm Security Administration photographer Russell Lee captured the scene of a neatly dressed man reading the paper, his honey stacked in orderly rows. Under the direction of Roy Stryker, the FSA photographers educated people about poverty and documented the changing agrarian lifestyles.

As has been the case since colonial times, women formed their social bees, especially during the darker days of the 1930s. The Busy Bees Social Club in Mitchell, Nebraska, was formed in the 1920s, but they stayed together until 1988. According to scholar Carolyn O'Bagy Davis, "their

5.15. Busy Bee Home Social Club, Louisiana, 1940. Courtesy of the Library of Congress, photo by Marion Post. Nurse White writes down an agenda, smiling in anticipation of another meeting.

minutes reveal how these clubs functioned as a vehicle for rural women to serve the local charitable needs, as well as respond to national events, such as the Depression and World War II."[116] The Busy Bees were quilters, but its real star was a woman named Fannie Schumacher, who not only organized the group, but kept the records and passed them on to her granddaughter, Arlene Buffington. These records give us a powerful insight into how the "bee" responded with optimism, a collective social conscious, and creativity to the nation as it defined itself in the twentieth century. The rules: everyone took a turn at hosting the club, even those who lived in sod houses. The hostess had to serve a lunch consisting of "two eats, drink, and pickles."[117] The Busy Bees even had a dress code: "Any member who comes dressed in anything but their house dress will be fined."[118]

The Busy Bees charged a dollar for dues, but that amount was reduced to fifty cents in 1938, and women could take the entire year to pay. Davis is also careful to mention that no record exists that a woman was dropped for failure to pay the dues. In 1932, the group had a Hard Time Party, where they agreed to work on a quilt that would be given to a family in need.

Farm Security Administration photographer Marion Post recorded the intimacy and camaraderie that existed among the Busy Bee Home Social Club in Thomaston, Louisiana. Nurse Marguerite White just happened to be making plans with Rosie Patterson, assistant secretary, when Post strolled by, taking pictures for Roy Stryker's Louisiana Delta project. Post's photographs illustrate the social cohesion that formed in the midst of crisis.

The 1940s: World War II

By the 1940s, Americans had emerged from the Great Depression and were producing as much honey and beeswax as possible for the war effort. World War II provided a major catalyst in large-scale honey and beeswax production. As in previous wars, military forces needed beeswax to waterproof canvas tents, belts, and metal casings of bullets. But beeswax made life more comfortable in the service in little ways as well. Writer Mary Kay Franklin explains: "Beeswax was used in adhesive tape. It was used as waterproofing and protective coating for shells, belts, coils and machinery, especially in warm climates where grease would run off. Beeswax did not mildew and was preferred over paraffin for waterproofing canvas. Large quantities were used in work plants for waxing cables and pulleys. Soldiers and sailors needed it for their dental work. Beeswax in skin protective sun creams and camouflage makeup for commandos was essential to the war effort. Airplanes waxed smooth with beeswax saved thousands of gallons of valuable fuel."[119]

In a 1942 article written by G. H. Cale, then editor of *American Bee Journal,* he quoted an official in the Office of Price Administration who calculated that beeswax had approximately 350 uses in the army and navy, and that there were 150 uses in the pharmaceutical fields.[120] In the article, Cale also explains that the normal imports from Brazil were suspended because the supply ships had been sunk. Thus, given that there would be beeswax shortages, Cale gave hints to beekeepers so they could increase their beeswax production. One simple trick was to remove a frame from a super, thus forcing the bees to build bigger, fatter combs on the frames. He also exhorted beekeepers to save every bit of beeswax, for "producing more beeswax—saving more beeswax—is a patriotic effort. It is part of the campaign for VICTORY. And more than that, it is your duty."[121]

The winter of 1942–1943 had been very cold, and because many beekeepers had taken so much honey the previous autumn, their bees died. So beeswax was even more valued in 1943. According to a *Newsweek* article, "War Food Administration urged beekeepers to conserve every ounce of beeswax, pointing out that more than a million pounds were needed this year for use in war products."[122]

Beekeepers served in the military in massive numbers and in a variety of ways. The future editor of *Bee Culture,* Larry Goltz, kept bee colonies until he was activated. When he came back, the colonies were gone![123] Russell Mitchell, who carried on his father's beekeeping business, served in the navy. William Rowley Jones came home from his service with the 507th Engineering Battalion, stationed in Alaska and Okinawa, to run his own bee business in Utah.[124] One of the Hispanic beekeepers, Felipe Garcia, served not only in World War II, but then again in the Korean War.[125]

The air force contained quite a few men important to the beekeeping industry. Eugene Killion was managing more than a thousand hives when he enlisted in the air force in 1941.[126] Floyd Moeller served as a navigator in the Army Air Corps, finally leaving as a lieutenant colonel before he returned to the bees.[127] Moeller later taught apiculture at the University of Wisconsin and was a researcher for the USDA. Harry Rodenberg of Wolf Point, Montana, served as a bombardier, achieving the rank of lieutenant before heading back to his commercial beekeeping business. Two brothers—Fremont and Larry—volunteered for the air force and the army, respectively. Fremont was shot down and spent three days in the English Channel before being rescued. Keeping bees probably didn't seem like such a bad profession after that experience.

Norman Sharp decided to use his enlistment as a form of professional development. While serving three years in the air force, Norman decided to hive the Giant Bee in India. In order to place Sharp's experi-

5.16. World War II bees. Courtesy of *Bee Culture.* This soldier shows off the headquarters for some of the war's busiest recruits. The military continues to use bees to gather information about chemical weapons.

ment in proper perspective, we must read Wyatt Mangum, who wrote in the *American Bee Journal* as recently as 2004: "Imagine a honey bee much larger than ours, whose worker bees look like our queen bees. . . . Imagine a colony of these bees living on one large comb, about half as big as a door, built in the open air and suspended from a large tree branch. In addition to foraging during the day, imagine these bees foraging at night. And now, to really strain the imagination, imagine these bees so defensive that they make Africanized bees, the so-called 'killer bees' seem mild by comparison."[128] The result at Sharp's air force base is easy to imagine: "The bees scattered all over and drove everyone out of the hangar," wrote Moffett, censoring the soldiers' language that must have been used in such an experiment. Moffett concludes: "One hanger had five swarms."[129] Even Harry Laidlaw served as an army entomologist, and one soldier refused to leave the bees behind. The First Battalion proudly welcomed all bees to its "Bee Platoon." There was even a water dish.

Some beekeepers chose to stay stateside. According to Franklin, "In 1942, the Selective Service System included beekeeping as an essential agricultural activity allowing local draft boards to grant deferments to individuals whom they felt contributed more to the effort through their beekeeping activities."[130] The Honey Industry Council persuaded the federal government to exempt wood, sugar, and metal from rationing for beekeepers because beekeepers supplied beeswax and honey to the government. Sugar was needed to feed the bees, and lumber was needed to make hives. Anything was done to ensure a steady supply of beeswax. In fact, at the government's request, Leon Winegar quit his job working at a Detroit auto plant in order to buy more colonies to produce beeswax.[131] In his 1941 catalog, Walter Kelley advised his customers of the following: "Every beekeeper is going to be called on to produce a maximum crop. To do this, you should melt up those old drone combs and replace every old queen. . . . Give each colony a weekly inspection . . . and above all give them plenty of room and practice swarm prevention."[132] And Kelley was not finished. He attached a string to the catalog and gave his customers strict instructions to hang the catalog on a nail by the wall because a new one would not be published during the war.

Sugar was rationed during World War II, so a beekeeper was an important person in many communities. Serving as a conscientious objector was beekeeper Oliver Petty, who spent summers smoke jumping in Missoula, Montana, and hiving swarms near Glenora, California.[133] Bonnie Sparks, a volunteer at the Daughters of the Utah Pioneers museum, could recall her mother sending her down to get honey from the local beekeeper during the war. "We would stand from a-way back," she said, eyes twinkling, "and watch the bees buzzing around him. We'd eat that honey right off the plate. Comb too."[134]

Perhaps because beekeeping was so important during the war, beekeeping magazines were not suspended as they had been during the Civil War. In fact, the normally frugal Walter Kelley took over publishing the regional *The Beekeepers Item* in 1944. Perhaps because of publishing restrictions, he wrote that the magazine, now to be called *Modern Beekeeping,* "combined with *Beekeeper's Review, Domestic Beekeeper, Dixie Beekeeper, Southern Bees, Western Honey Bee,* and *Bees and Honey* and superseding *The Beekeepers Item.*"

But the War Procurement Board, which rationed sugar and lumber, could be a formidable foe. Occasionally, a War Procurement Board member would disagree with a bee inspector, who had even more responsibility to ensure hives stayed healthy during the war. In one story, inspector John Bruce had given a beekeeper permission to buy sugar to feed his bees, but the War Procurement Board secretary vetoed the beekeeper's request. As a result, the bees were starving.

The War Board woman was not sympathetic to the beekeeper, asking him: "Who do you think you are, trying to get sugar for his bees when I can't get sugar for my coffee?"[135] When the beekeeper returned with John Bruce in tow, the woman relented under threat of a federal lawsuit. The beekeeper finally was able to feed his bees, but most of his profit that year literally died with his bees.

Occasionally, beekeepers could use a good cup of coffee to get on the good side of an inspector. The most delightful honey house inspector story from the 1940s involves Willie the Enforcer. Clay Tontz writes about the post–World War II days in Covina, California. Despite his short stature, Inspector Willie could strike fear in the hearts of honey sellers: "He was all in black—black suit, black felt hat, the brim turned down over dark eyes, gangster fashion. In his right hand he carried a huge, black fountain pen, gripped and pointing forward like a gat. Under his left arm he carried a long, narrow record book (black, of course) about as long as his arm." Tontz had one ace—his wife, Jeanette. While Willie the Enforcer was busy writing his list of violations, Jeanette applied her makeup, started the coffee, and pulled out apricot pie. When she insisted he join them for dessert, Willie the Enforcer met his match. "The great, menacing black fountain pen faltered, then stopped," writes Tontz. "His long book clamped shut like a pair of crocodile's jaws at lunch."[136]

Dances with Bees

Amidst the massive carnage of World War II, Karl von Frisch published *Bee Dancing and Communication* (1944), his findings about bee communication and social order. He was not the first to theorize that bees could communicate. Indeed, the early Greek writer Aristotle noticed that recruits were sent out in search of food. In *The World History of Beekeeping and Honey Hunting,* Eva Crane includes a quote from book 9 of the *Historia animalium* describing the foragers: "'On reaching the hive they throw off their load, and each bee on his return is accompanied by three or four companions. One cannot tell what is the substance they gather, nor the exact process of their work."[137]

As late as the twentieth century, writers assumed that bees were dancing for joy because they had found pollen. Crane includes a quote from E. R. Root (son of A. I. Root) writing in 1908 that a dance was "a sign of joy that they have found pollen, and that they take this means to communicate the knowledge of it to their fellows."[138] The black researcher Charles Henry Turner conducted color and pattern experiments in St. Louis in 1910, but his findings were overlooked. O. W. Park successfully tracked the dances of water-collecting bees, and these findings established the foundation from which von Frisch drew for his studies. So obviously, bee researchers had been interested in dances for a long time. But von Frisch was the first one to clearly describe and illustrate these dances in ways that a general audience could understand.

In order to communicate pollen and nectar sources, honey bees developed two dances: the round and the waggle dances. When a forager honey bee returns to attract recruits to the new sources, it can use these two dances to communicate distance and direction of the flowers. A round dance is performed when food is within ninety meters of the hive. First, the forager bee unloads its supply to its waiting sisters. When it is on the comb, it begins the dance: "On the part of the comb where she is sitting she starts whirling around in a narrow circle, constantly changing her direction, turning now right, now left, dancing clockwise and anticlockwise in quick succession, describing between one and two circles in each direction."[139] But the other bees in the hive will join the forager bee: "What makes it so particularly striking and attractive is the way it infects the surrounding bees . . . those sitting next to the dancer start tripping after her, always trying to keep their feelers on her," explains von Frisch. In this way, the other bees can memorize the directions and go scout for the sources themselves.

The waggle dance is performed when food sources are farther than ninety meters away. The forager bees then must convey more information to their sister-recruits. The flower's relation to the sun is more important in the waggle dance. Says Frisch: "The characteristic feature which distinguishes this 'wagging dance' from the round dance is a striking, rapid wagging of the bee's abdomen performed only during her straight run."[140] The number of waggles increases with distance. Thus, the more waggles a dance has, the longer the distance the recruits must fly to reach the source.

James and Carol Gould best explain how the bees learn direction: "When the waggle run in a dance is pointed up (the dances are performed on vertical sheets of comb), the feeding station is always in line with the sun; when the food source is directly away from the sun as viewed from the hive, the dances point down; when the food is located 80 degrees to the left of the sun, the dance points 80 degrees to the left of vertical. A bee attending a dance need only determine the orientation and duration of the waggle run in order to know the distance and direction of the food."[141] These dances provide much-needed communication between the forager bees and the recruiter bees. Von Frisch was awarded the Nobel Prize in 1973 for this insight into honey bee communication.

Early in her life, the queen leaves the hive, although it is not to forage for food. In order to mate with the drones, the queen goes on several nuptial flights. The queen generally mates with the drones that fly the highest and are able to catch her. In unsympathetic language, Sue Hubbell describes the process, "When a drone sees a queen, he flies high in the air to mate with her. He mates by inserting his penis into her sting chamber, which closes around it, causing it to rip loose from his body as he bends over backward and falls lifeless to the ground."[142] The queen will mate a number of times in the air and then return to the hive. Beekeepers check to see if the queen has been out by looking for the drone's entrails, something that they call the "mating sign."

When the artificial insemination industry was in full swing, essayist E. B. White wrote a hilarious poem satirizing the human's attempt to control the bee society, especially the queen. Published in the *New Yorker* in 1945, "Song of the Queen Bee" is told from the queen's perspective and protests the industrialization of reproduction:[143]

Song of the Queen Bee

"The breeding of the bee," says a United States Department of Agriculture bulletin on artificial insemination, "has always been handicapped by the fact that the queen mates in the air with whatever drone she encounters."

When the air is wine and the wind is free
and the morning sits on the lovely lea
and sunlight ripples on every tree
Then love-in-air is the thing for me
    I'm a bee,
    I'm a ravishing, rollicking, young queen bee,
    That's me.
I wish to state that I think it's great,
Oh, it's simply rare in the upper air,
    It's the place to pair
    With a bee.

. . .

There's a kind of a wild and glad elation
In the natural way of insemination;
Who thinks that love is a handicap
Is a fuddydud and a common sap,
For I am a queen and I am a bee,
I'm devil-may-care and I'm fancy-free,
The test tube doesn't appeal to me,
    Not me,
    I'm a bee.
And I'm here to state that I'll always mate
    With whatever drone I encounter.

. . .

Mares and cows, by calculating,
Improve themselves with loveless mating,
Let groundlings breed in the modern fashion,
I'll stick to the air and the grand old passion;
I may be small and I'm just a bee
But I won't have science improving me,
    Not me,
    I'm a bee.
On a day that's fair with a wind that's free,
Any old drone is a lad for me.
I've no flair for love moderne,
It's far too studied, far too stern,
I'm just a bee—I'm wild, I'm free,

That's me.
I can't afford to be too choosy;
In every queen there's a touch of floozy,
    And it's simply rare
    In the upper air
    And I wish to state
    That I'll always mate
With whatever drone I encounter.

. . .

If any old farmer can keep and hive me,
Then any old drone may catch and wife me;
I'm sorry for creatures who cannot pair
On a gorgeous day in the upper air,
I'm sorry for cows that have to boast
Of affairs they've had by parcel post,
I'm sorry for a man with his plots and guile,
His test-tube manner, his test-tube smile;
I'll multiply and I'll increase
As I always have—by mere caprice;
For I am a queen and I am a bee,
I'm devil-may-care and I'm fancy-free,
Love-in-air is the thing for me,
    Oh, it's simply rare
    In the beautiful air,
    And I wish to state
    That I'll always mate
With whatever drone I encounter.
All hail the queen!

Sex was indeed in the air for war-weary Americans, and Delta blues singers appropriated the honey bee in their songs about romantic relationships. The blues singers knew about difficult relationships: they were black in a white world. The combination of the honey bee image, bottleneck slide guitars, and a gravely voice worked its magic with at least two musicians: Muddy Waters and Slim Harpo. The blues—well, the blues made listeners feel better about life's bitterness, or at least less alone.

If you were black and wanted to sing in Mississippi, you could play in a juke joint almost any night of the week somewhere on Highway 61. But if you were black and wanted to get paid real money, you headed up to Southside Chicago, which joined New York as a center of black culture in the 1930s. It offered more freedom, more economic opportunities, and more music. By 1930, the largest population of Mississippians outside the state was in Chicago.[144] As America's entrance into World War II stoked the industrial fires of the northern factories, the need for soldiers created a labor crisis in the factories. Of the African Americans who went north in the first half of the century, nearly half migrated between 1940 and 1947.[145]

A cotton sharecropper named McKinley "Muddy Waters" Morganfield had a habit of composing songs to pass the time while riding on a tractor in Mississippi. One of his most famous, "Honey Bee," became a hit after he moved to Chicago in 1939. Although blues musicians didn't hit the big time until the 1950s, the solid black cultural base began to form in the cities as early as 1915 with Jelly Roll Morton's first jazz notations and in the 1920s with Mamie Smith's "Crazy Blues."[146] During the 1940s, when the black veterans returned from World War II and relocated from the rural areas to the urban cities, the blues found a new energy. When black musicians started singing about honey bees, they were singing to an audience divorced from the land, their family, and their heritage.

Postwar Progress

After the war, Congress realized the need to provide and protect honey bees in order to maintain the food supply for this country and other nations if it wanted to attain superpower status. In 1949, Congress passed the first price support programs for commercial beekeepers. The government's aid was long overdue, but the price support program had its own flaws: Congress only took honey and beeswax economic values into consideration and completely ignored the value of pollination services. Furthermore, many beekeepers unloaded their inferior grades of honey onto the government. Still, the price support program offered stability to honey producers for the first time since it had become an industry.[147]

Research funds were also appropriated to discover quick, efficient chemical ways to combat age-old beekeeping problems. The knowledge of sulfa drugs learned in World War II was transferred to the beehive: Leonard Haseman and L. F. Childers learned that sulfa drugs could contain the spread of AFB. AFB is a bacterium caused by *Paenibacillus larvae,* formerly called *Bacillus larvae.* The compound sodium sulfathiazole provided a short-term control by suppressing the symptoms. It also prevented the reproductive spores from germinating. Consequently, large-scale commercial beekeepers got a boost.[148] Administering sulfa as a powder or mixed in sugar syrup did not require the immense amount of labor that shaking bees onto new foundation demanded, or the expensive practice of killing colonies and burning the contaminated equipment. Researcher Bill Wilson explains why commercial beekeepers adopted the chemical cure: "Eventually most beekeepers and inspectors realized the economic value of protecting colonies with the chemical treatment. It wasn't long before the practice became widespread. From that point on, the ancient scourge of beekeeping, foulbrood disease, lost much of its impact."[149] The availability of sulfa drugs had a downside: research about natural bee resistance to AFB was neglected until the latter part of the twentieth century. Thus, O. W. Park's studies, which had been done in the 1930s, were shelved until the 1980s.

The buildup of the United States as a cold war superpower required the participation of bees. In order to test chemical warfare, bees were used to check radioactive pollen from land areas where tests had been done. The army used the Aberdeen Proving Ground in Maryland as a dump for chemical warfare agents, unexploded munitions, and wastes from research and production facilities throughout the 1940s. Because seepage from these materials affected surrounding insect life significantly, the army still maintains wired hives with infrared bee counters to track flight activity and search for deviations from the insects' normal patterns.[150]

Even as the United States was looking into the future to define new strategies for war, political power, and diplomacy, many communities were looking backward to celebrate the past. Salt Lake City commemorated the honey bee in its 1947 centennial parade. On one massive float, one lady stood out above all the rest, with a huge crimson hoop skirt, gold hives sewn around the hoop, and gold bees on the gossamer fabric sleeves. The other ladies, sporting dresses with wagon wheels and seagulls, did not stand a chance next to the queen bee.

This beehive dress embodied a classic dilemma facing 1950s America:

5.17. Salt Lake City Mormon celebration float, 1947. Courtesy of Utah State Historical Society. Each gown featured in this photo is an important symbol to Mormon society—the bee skep, the wagon wheel, and the seagull. The bee skep dress is on the right. The dress itself is crimson with gold skeps. Bees are on the sleeves.

the desire to move forward, but the simultaneous desire to hold to symbols of the past. In the Jerome Lawrence and Robert E. Lee screenplay *Inherit the Wind,* the lead defense attorney summed up the effects of technology on American culture in his closing argument: "Progress has never been a bargain. You've got to pay for it. Sometimes I think there's a man behind a counter who says, 'All right, you can have a telephone, but you'll have to give up privacy, the charm of distance. Madam, you may vote; but at a price; you lose the right to retreat behind a powder-puff or a petticoat. Mister, you may conquer the air; but the birds will lose their wonder, and the clouds will smell of gasoline!'"[151] Scientific advancements and improved transportation further opened the possibilities of a commercial honey industry, but in doing so, beekeeping lost the appeal of a cottage industry. With the opportunities for federal support came the need for legislation and regulation. When people moved to the cities, they forgot the joy of opening a hive to get honey. When they preferred electricity, they forgot the pleasant scent of a beeswax candle. So they recaptured those pastoral ideals by turning to the arts—to songs, to books, to poems, and to ballroom dresses—without having to forego the hard-won accomplishments earned during the Depression and two world wars.

Chapter 6

Late Twentieth Century
Globalization, 1950–2000

> Nature has unlimited time in which to travel along tortuous paths to an unknown destination. The mind of man is too feeble to discern whence or whither the path runs and has to be content if it can discern only portions of the track, however small.
>
> —*Karl von Frisch*

Almond trees dot the California valley like Degas ballerinas. The limbs, pruned precisely so that sunlight will strike the middle of the trunks, fan out like pink tutus. The blossoms have a dusky, dusty scent so thick it is almost cloying. Almost. Little did Franciscan priest Junipero Serra know how the almond trees he brought with him to California in 1767 would affect American beekeeping. It is not by accident that the Spanish would transplant so many crops to California. The state has many of the same Mediterranean conditions that defined successful Spanish agriculture—that is, approximately fifteen to twenty inches of rain a year, dry heat, good drainage, and chilly winter temperatures. Before Father Serra died in 1784, he had established nine missions along the western branch of El

Camino Real, which stretched from California to San Antonio. In the 1840s, growers were finally experimenting with hybrids and grafting and each mission contained a grove, although the perfect tree eluded them.

By the 1950s, orchard growers finally accomplished what they were striving for: a good combination between an irrigated valley floor, almonds grafted to a peach root stock, and acreage—a lot of acreage. Half a million acres, to be exact. More acreage than Spain, the world's primary almond supplier until the 1950s, to emphasize the point. In short, California provided ideal growing conditions for producing bumper crops of almonds for the world—if beekeepers could provide the bees to pollinate the almond orchards. Almond trees are not self-pollinating, and because of this basic fact, the ancient relationship between bees and cows shifted subtly in twentieth-century America. Although it had been proven that cows and cattle could succeed virtually anywhere in America, almonds most assuredly could not. If specialization was the key to succeeding in the industrial countryside, almond growers captured the moment in the 1950s, and beekeepers were an inextricable element in their formula.

When America redefined its position as a major power in world politics after World War II, the country also had to redefine its relationship with beekeepers. If indeed America would become what Steven Stoll, in *Fruits of Natural Advantage: Making the Industrial Countryside in California,* calls the industrial countryside, the country would need beekeepers to provide pollination not just to almond groves, but to pumpkins, blueberries, cucumbers, cherries, and strawberries. Recognizing in a small way the importance of bees, the federal government provided beekeepers with more assistance and protection than they had in any other century of American history. When America became a world supplier of food, pollination became a major way for beekeepers to supplement their incomes. However, during this period of globalization, American beekeepers would have to redefine how they operated given the number of pesticides, new imports, land zoning regulations, federal quarantines, and chemicals.

Technology affected how beekeepers approached their business. Airplanes were both a help and hindrance: beekeepers could scout new fields, but pesticide applicators could drop chemicals whenever and wherever the winds would scatter the load. Forklifts, hive lift machines, honey pumps, and semis would become the norm by the end of this century. Improved interstates provided the lifelines to the migratory beekeepers.

Culturally, technology affected how the honey bee was perceived by the public. Television, computers, and the Internet became powerful forces in transferring information. Ironically, the century noted for commercialization, federal legislation, and scientific research ended with more questions than answers about bees. Never before had there been so many possibilities to inform or misinform the public about beekeeping. Although the media hype about the African honey bee created a near hysteria until the 1980s, other forces were at work to educate and entertain the public about not only the African honey bee but also mites, imports, and pollination.

Beekeepers still struggled for a variety of social, ecological, and financial reasons, however. Commercial pesticides were not well regulated, so bees died at alarming rates. Honey prices dropped after World War II with the easy availability of sugar, and "heavy outbreaks of European foulbrood (which did not have a chemical treatment) caused beekeepers to go nearly bankrupt," according to American foulbrood researcher Bill Wilson.[1] In addition to European and American foulbrood, other diseases and pests such as chalkbrood (1972), tracheal mites (1984), varroa mites (1987), and small hive beetles (1998) caused significant turmoil in the fields and the pocketbooks of beekeepers. By the 1990s, the African honey bee suddenly seemed a very small threat, if only because it showed some resistance to the varroa mite.

Indirectly, the postwar housing shortage created a building boom of suburbs, representing more long-term threats than people could possibly realize when William Levitt built his first town in Hempstead, New York. By converting farmland into premanufactured houses, Levitt could build an unspecified number of houses within months. According to David Halberstam, "Starting in 1950 and continuing for the next thirty years, eighteen of the nation's top twenty-five cities lost population. At the same time, the suburbs gained 60 million people. Some 83 percent of the nation's growth was to take place in the suburbs." In fact, Halberstam goes on to state: "Bill Levitt had helped begin a revolution—that of the new, mass suburban developments. . . . By 1955 Levitt-type subdivisions represented 75 percent of the new housing starts."[2]

Bill Levitt was not alone. Government loans and GI benefits fueled this massive building boom. Very few restrictions were placed on land usage. Suddenly, the dream of owning a home was not impossible, especially if the buyer was white. All of these elements spelled trouble for beekeepers fighting a constant battle against pesticides, lost forage area, and zoning restrictions.

Furthermore, because major social rifts were threatening to tear apart the "good life" that many Americans had thought they created, this country's arts environment used the honey bee to negotiate difficult power struggles between races, between spouses, between political parties, between generations, between legal rulings. The entertainment industry was the most visible force in redefining the honey bee's new values. Because of successful black musicians such as Muddy Waters and Slim Harpo, the honey bee became synonymous with the pains and frustrations associated with love and intimacy. The blues singers had a short period of popularity compared with rock 'n' roll musicians, but they left an indelible impression on the late twentieth century. Likewise, the gospel group Sweet Honey in the Rock used the image of honey to sing against political and gender oppression during the civil rights movement; the group's legacy has extended beyond the 1960s.

Photographers and movie directors recorded the daily details that feature honey bees or the people who work with them. Photographers such as Richard Avedon redefined the bee hunter in the twentieth century: he or she is no longer on the outskirts of society, but someone who looks like the next-door neighbor. He and other photographers detailed the difficulties of survival in this new land of milk and honey.

For the first time in the history of the honey bee as a cultural symbol, the bee was a villain in *The Swarm,* reflecting the country's fear and paranoia of the African honey bee. However, by the 1980s and 1990s, the country had regained some of its equilibrium. *Fried Green Tomatoes* (1991) offered a female bee charmer on the silver screen. Dramatizing the link between military service and beekeeping for the first time, movie director Victor Nunez used the honey bee as an appropriate reconciliation metaphor in *Ulee's Gold* in the 1990s.

It's no surprise, however, that baseball and finally softball teams continued to adopt the honey bee as their mascot after World War II. With the issue of race becoming an explosive issue, sports and schools became the avenues for the American public to address racial and gender prejudice. Because Americans don't have a national religion, sports provide a way for people to share rules and values. From the Burlington Bees to the Salt Lake Buzz, baseball teams chose the honey bee as their icon because such a symbol emphasized a tightly organized social infrastructure, which good baseball teams need.

During the late twentieth century, the honey bee reflected a wide range of values in a variety of media, complementing the complexity of American domestic and international policies. By the end of the 1990s, the honey bee no longer just meant industry and thrift or even a central organized society. The many cultural contexts in which the honey bee appeared reflected how much the American Dream always had been an interdependent global dream.

The 1950s

In an effort to bring stabilization and control to wildly fluctuating honey-market prices, Congress voted in 1949 to pass a price support system that would encourage small farmers to become honey producers. According to Pamela Moore, the Honey Support Program (1950) encouraged beekeepers to remain in business so that pollination of commercial crops would be ensured.[3] The benefits were good because beekeepers had a ready market and did not have to contend with wildly fluctuating prices. Put simply, according to Douglas Whynott, "A beekeeper used the honey support program by taking a loan and using the honey as collateral. The loan allowed him to pay operating costs, and to hold the honey until he found the best market. After the sale, he paid off the loan."[4] The Commodity Credit Corporation (CCC) acted as the middle agent between the government and the beekeeper. During the first two decades of the price support program, the market price was higher than the price the government offered beekeepers.

Still, afraid that a cycle of dependency would be nurtured between beekeepers and the government, Walter Kelley cautioned beekeepers, "government support should be looked at as only temporary."[5] Later in the year, he urged beekeepers in 1953 to buy back the surplus honey from the CCC offices located in different regions. He stated, "By so doing you will show the government that you are in earnest in doing your best to make this price support program work and at the same time you will be holding your trade and making some profit."[6]

However, American foulbrood remained the nemesis it had always been because beekeepers could not administer Terramycin as quickly and efficiently as it was needed in the hives. In fact, *Bee Culture* writer Karl Showler declared that the industry almost bottomed out. Researcher Bill Wilson explains the beekeeper's dilemma: "Bee researchers determined that diseased colonies needed to be dusted with Terramycin on a weekly schedule for five or more times (200 mg of oxytetracycline each dusting with a total of 1,000 mg or more per treatment period per colony). This schedule was difficult for commercial beekeepers with widely scattered apiaries. Some bee yards were more than 100 miles from the center of the beekeeper's operation. Travel costs were high."[7] When Wilson devised sugar Terramycin patties (which could be left in hives and slowly consumed by the bees over time) in the late 1950s, beekeepers could medicate their hives in larger intervals, and American foulbrood ceased to be as great of a threat.

With the addition of Alaska and Hawaii as states, westward migration came to an end, although the two states could not have been more different. Honey bees had been in Hawaii since the 1800s when Catholic missionaries arrived in the islands. Hawaiian beekeeping followed the cattle industry, as it had in Texas and California. So when Catholic missionaries arrived in Hawaii, they brought the descendants of the Texas cattle that had first been brought from Spain in the 1580s. The missionaries also introduced the mesquite tree, which is known as *kiawe* in Hawaii. With mesquite trees, cattle, and bees, many Hawaiian plantations flourished during the early part of the century, although honey was considered a secondary interest compared to the sugarcane industry.

However, Hawaiian bees declined in the 1950s. "The stimulus of WWII was insufficient to rebuild the industry to its former proportions," according to J. E. Eckert, "and in some instances the number of colonies and production dropped still lower."[8] The Hawaiian beekeepers had difficulties with mismanagement, American foulbrood on some islands, and the disruption in trade as a result of World War II. Compared with the honey industry in the 1900s, the post–World War II Hawaiian honey industry was tepid at best.

Alaskan beekeepers were just getting started, on the other hand. Starting in the 1920s, beekeepers had two obvious challenges: difficult winters and transportation of package bees. These challenges overshadowed the "normal" challenges of forage plants or labor concerns that other states may have, according to Stephen Petersen.[9] The winters routinely dipped to -45°F. Transportation prices to Alaska were high, so a full-scale industry developed very slowly.

Not true in California. It was there in 1951 that "my dad [C.F.] cashed his first check for pollinating almond orchards," reminisced honey bee package producer Bob Koehnen.[10] The Almond Board of California, created in 1950 by the U.S. Congress, promoted the relationship between almonds and honey bees. With extensive research paid for by almond companies, growers were able to learn the optimal grid patterns for orchards and the formulas of hives per acre that would ensure adequate pollination of their almond trees. For every ninety-five trees, two hives should be planted to ensure maximum pollination. Every forty years, the almond trees should be dug up and replanted with fresh stock. The result of these studies is easy to see: More than six thousand almond growers farm over half a million acres in California.[11] California is the only place where almonds are commercially grown in America, and it also supplies 80 percent of the world's almond needs.[12]

As far as beekeepers are concerned, almond orchards provide one of the earliest sources of nectar and pollen, which is important to commercial beekeepers in need of colony buildup. Because almond trees are not self-pollinated, beekeepers can rent their hives for higher prices between February and March to almond growers. Writing in 2000, Roger Morse explains the interconnected relationship between beekeepers and orchard growers: "More colonies of honey bees are owned and operated in California than in any other state. California usually leads the nation in honey production. In the 1989 paper, it was estimated that 70 percent of the colonies of honey bees rented for pollination were rented in that state, where, of course, more colonies were used for almond pollination than for the pollination of any other single crop."[13]

Although there is great variation between pollination schedules and crop varieties, extension specialist Eric Mussen suggests that the 1950s were when the research studies about pollination finally started to transfer to a practical application in the fields.[14]

Commercial pollination was happening in other states as well: wild blueberries in Maine, apples in Washington, cherries in Michigan, cucumbers in Ohio, Madrid sweet clover in Texas. However, given the unique growing conditions in California, almond growers have had a niche on that market. "If you want to see supply-and-demand economics," explains Eric Mussen, "this is it."[15]

Another factor in this successful agricultural formula, then, is the queen and package bee producer. The queen-rearing industry in California became more integrated with the almond industry in the 1950s. Bob Koehnen explains that package bee producers have had two interrelated responsibilities over the years: to keep the bee lines as true as possible, and then use them as good crosses when bee breeding so that good traits are passed on to the young bees.

When preparing his packages, Koehnen says nothing is left to chance. Young colonies are set in the almond orchards because almonds generally are the first crop to bloom. Thus, foragers can bring back nectar and pollen, allowing colonies to gain strength and numbers. Later, they produce the queens developed for packaged bees. It is a case where "the trees take care of the bees, and then the bees take care of the trees," says Bob's wife, Yvonne.[16]

The package industry is labor intensive. For a hundred days, seven days a week, Koehnen's crew works at grafting queen cells with larvae, filling nucleus boxes called cell builders, and driving these cell builders into orchards. Twelve people are queen cagers; these people place the young queens in a small wooden shipping cage that has enough sugar to sustain the queen while the package is being shipped to a beekeeper.

The cell builders are brought back from the fields as strong colonies, the queens are caged and ready to be packaged, and an assembly line creates a mail-order package to send to beekeepers. The Koehnens have worked out a system with the United Postal Service that ensures efficiency in mailing. The Koehnens ship their queens the same day the queens are caged and put in a package. His company has achieved a "marriage of quantity and quality on an efficient schedule."[17]

Just as pollination was beginning to be big business, honey prices slumped after World War II, aggravated by the unregulated use of pesticides. Beekeepers had learned that migratory beekeeping was the most profitable way to stay in the industry: "Migratory beekeeping had become the rule rather than the exception," according to Charles Lesher.[18]

Owners were required to register locations, and their routes crisscrossed the United States from as far away as Florida, Texas, Arizona, and California to Washington and North Dakota. But because beekeepers were often the last to know about pesticide schedules, they incurred heavy losses. Chemicals didn't affect other farmers' livestock or crops, so beekeepers often were left with little recourse, except through a federal compensation program.

If chemicals didn't affect schedules, weather did. Writing of the Rio Grande beekeeping routes, Clay Eppley details a series of weather-related routes that completely disrupted migratory routes to that region. In 1949, there was a killing frost. Then in 1950, a torrential rainstorm and even boll weevils plagued beekeepers. Finally, in 1951, Eppley described the "daddy of all Valley freezes . . . and in only a few hours our Valley citrus orchards were gone and with it our fine crops of citrus honey."[19]

But beekeepers weren't the only ones following migration routes.

Los Abejas Bravados

Initially, the African honey bee bore the burden of bad press and bad timing. Dubbed the "killer bee" by the American media, this honey bee was unfairly vilified, as was the researcher who studied it.

Dr. Warwick Kerr, the great-grandson of a Civil War Confederate soldier, is a renowned geneticist and specialist in native stingless bees. He is also an advocate for the poor and oppressed in Brazil. In 1956, Kerr won a cash award from the Brazilian government that would combine both of his interests: studying bees and helping poor people. The award funded projects that would help Brazil's poor people learn skills. Beekeeping is a low-cost agricultural enterprise to initiate compared to other industries because beekeeers do not need land to operate. Kerr thought beekeeping would provide a good sustainable economy if Brazil had a bee better adapted to the tropical region. African bees, despite their temper, have a great gift: they are survivors. The African bee is, to quote researcher Marla Spivak, "well-adapted to subsistence living. It can find patches of nectar that other honey bees will overlook."[20] African bees put all their energies into population building. Thus, they tend to swarm more often than Italian honey bees, and when their nests are threatened, they are more defensive than other honey bees. The initial hope that the African bees could be hybridized with other strains did not happen quickly, and it remains a question whether hybridization occurred at all during the twentieth century.[21]

Forty-seven African queen bees survived introduction to São Paulo University, Brazil. But the bees were soon shipped to a eucalyptus forest owned by a railroad company. The accident occurred there, according to Dewey Caron: "A railroad employee removed the double queen excluders from the entrances of 26 of these colonies in early 1957, apparently because he saw pollen loads being removed from workers returning to the colonies. The colonies swarmed soon thereafter."[22] The African bee started swarming north, but the media moved even more quickly. "Killer bees" made for great headlines.

As if that misfortune were not enough, when Kerr criticized the Brazilian military government for its treatment of the poor, he was imprisoned. While imprisoned, the Brazilian government criticized Kerr's research with the African honey bees. Even though Brazil eventually became a large honey producer, Kerr was not appreciated by Brazil or America until the latter part of the twentieth century, when it became apparent that there were more devastating enemies to North American honey bees than African bees.

In the meantime, American beekeepers had their hands full with American and European foulbrood, pesticides, and even some infighting between beekeepers themselves, if H. L. Maxwell was correct in his assessment of beekeeping in 1959. Maxwell, using Virginia as an example, states that commercial beekeeping in Virginia was difficult. Commercial beekeepers tended to see small honey producers as competition, and the small honey producers saw the commercial beekeepers with a type of envy. He reported that meetings were rarely attended, and younger members were impossible to attract. Maxwell stated, "The most sustained effort at any official level has been that of the state inspector, and *it is essentially a police effort.*"[23] He suggested that the average beekeeper was still an independent character, and it was not unlikely for an inspector to be met at gunpoint.

"Independent" adequately describes one Alabama beekeeper named Euclid Rains, totally blind since a childhood accident scarred his eye tissue when he was three. His parents were determined not to treat him any differently, and so he grew up on the farm, orienting himself by using sound. In 1956, he bought a couple of hives from two men who delivered the bees, but then left. So Rains had to move the hives to a place where the bees would not be in the way. As he says, "The bees were very unhappy with the way I was handling their homecoming party and were more than vocal with their objections."[24] In 1957, he was the first person to receive a loan from the Farmer's Home Administration. He is reported to have said, "If I was not too blind to pay taxes, I was not too blind to borrow tax money."[25]

Indeed, beekeepers were still independent, even if the rest of the nation was watching television shows that advocated conformity, such as *Leave It to Beaver* and *I Love Lucy.* California beekeeper Clay Tontz describes a hermit beekeeper nicknamed "The Mean Scarecrow," who hung out in the ravines bordering Tontz's beehives. The hermit came from Tennessee and spoke with an accent. To defend his place, the Scarecrow had devised "a hundred sorry beehives . . . filled with attack bees (short-fused black bees). . . . Ropes were attached to the guard hives which were set on loose stones." If anyone were to visit without warning, the Scarecrow could rock the hives and thus set loose a storm of bees on the victim. The Scarecrow was self-sufficient, but the "fruit jars of sloshing liquid brought down from the hills to customers after dark and labeled 'honey' had a kick like nothing ever carried in by the most zealous bee," Tontz said diplomatically.[26]

The Scarecrow's days were numbered. He represented a time when challenges were easily confronted with a rifle, when boundary lines were clearly drawn—provided he was holding the rifle. But atomic weapons changed how the United States determined power. The hydrogen bomb fundamentally affected how a war would be fought, and the Korean War showed a clear need for diplomacy.

Because the United States was considered a police power during the Korean War, its role was murkier than during World War II. Although food supplies were rationed during World War II, there were no noticeable interruptions in food staples as a direct result of the Korean War. However, the federal government did control wages and bee supplies. Wages at atomic power plants were considerably higher than at other factories. Thus, bee supply companies had a difficult time finding a labor supply. Bee companies like Kelley's and Dadant's and Root Company had hired women much sooner in the twentieth century than other industries. But World War II created a labor shortage. To fill the void, many women entered the bee industry, and they stayed even after the war had ended. Kelley, using a commercial honey packer named Walter Diehnelt in Wisconsin as an example, strongly advocated that women be incorporated in the honey houses across the United States. In an article titled, "Why Not Use Girls In the Honey House?" Kelley vouched for women employees: "We find that the girls are very neat and clean in their work and that their writing is more legible."[27] By the time he had written this article, Kelley had given up trying to maintain a workforce in Paducah, Kentucky, which was building an atomic power plant and paying workers more money. He moved his bee supply company to Clarkson, Kentucky, because he knew he could cultivate a loyal workforce and have cheaper shipping costs.

Thus, the Korean War represents the beginning of the cold war era in which honey bees do not serve an obvious function in a military endeavor, although beekeepers certainly served, as they had done in previous American military operations. Norman Mitchell, for example, not only grew up a beekeeper in Idaho, he also came from a family of veterans: his father served in World War I and his brother in World War II. Norman served as an aviation engineer during the Korean War.[28] When he returned, Norman returned to beekeeping, although he and his brother moved the business to Montana. As previously mentioned, Felipe Garcia, having already served during World War II, went to Korea. In a different form of service, Howard Graff "packed most of the honey that went to the Army Quartermaster in Korea and other South Pacific areas."[29]

The Black Migration

An equally serious conflict was playing itself out in the United States in clubs, airwaves, and school playgrounds. Since slavery, the black population in the United States had been defined by migration and segregation, according to music historian Mike Rowe. When blacks began to leave the South in the 1940s, they took the most direct routes to the North, which were the railroad lines to St. Louis, New York, Detroit, and Chicago. By the 1950s, blacks would go west to Los Angeles as well.

Furthermore, according to Rowe, the migration was "two-dimensional," because blacks were moving away from farms into cities. The black communities were segregated in these cities, but the 1950s events would fundamentally alter race relations because, at least in the North, the black communities had achieved stability, if not equality. Once in these cities, blacks organized voting blocs, churches, and clubs.

The blacks brought agrarian images and southern music styles with them. In Chicago, one of the black newspapers was the *Bee*. This newspaper did a lot to promote black musicians, politicians, business owners, and social calendars. Black musicians sang about the honey bee from the South all the way to St. Louis, to New York, to Detroit, to Chicago, and later, Los Angeles. Because these newly formed urban communities still had links to the South, the urban blacks financed a healthy cultural climate in which musicians developed a rural-based blues genre. Radios offered blocks of time to black musicians, and many musicians were able to profit by this promotion. In their songs, black musicians challenged the Victorian pretenses of modesty and chastity that had been lingering since the Anthony Comstock laws of the 1880s.[30] The honey bee served as a metaphor to make bold pleas for love, sex, and affection. These images, combined with a heavy emphasis on the backbeat, fused with jazzier harmonic patterns for a cool new sound, as in a song by Slim Harpo, whose protagonist boldly declares: "I'm a King Bee, Baby / Won't you spread your love all over me?"

Harpo's sexually charged lyrics appealed to both black and white audiences. This song influenced the Rolling Stones, who sang a cover of "I'm a King Bee" when they first played in America. Mick Jagger, befuddled by the racism existing in American rock 'n' roll in 1968, asked his listeners at one concert, "Why do you want to hear us when you can listen to Slim Harpo?"[31] The answer was literally as plain as their noses on their Anglo-Saxon faces.

The Rolling Stones also listened to Muddy Waters.[32] On July 11, 1951, Muddy Waters recorded "Honey Bee," a standard that he had written earlier when he worked on the Stovall cotton plantation back in the 1930s. That song also inspired Eric Clapton in 1958; Clapton declared, "I knew I could play guitar if I had mastered Muddy Waters' 'Honey Bee.'"[33]

These blues players used the honey bee to sing about the heartbreak of unrequited love—or, just as often, about the joys of illicit love, either with a partner who is too young, too married, or too dangerous. According to Muddy Waters's biographer Walter Gordon, the songs were "about sex with someone else's wife, sex with someone else's girlfriend, sex and trouble. . . . sex was sex, but sex also became an analogy for a kind of freedom, a freedom to serve himself, to damn the torpedoes, the shift supervisor, and the overseer's big gun."[34] Postwar Chicago was the place for these displaced Southern men; "The Chicago Black Renaissance reflected the impact of [FDR's] Works Progress Administration on the lives of artists," according to historians Hine, Hine, and Harrold.[35] Chicago offered a huge recording industry for black singers. Fortunately, the British and Irish teenagers didn't have the same racial prejudices affecting American recording industries; they embraced Muddy Waters, Slim Harpo, and their aggressive bees, which captured the sweetness and the painfulness of life and love in the black world.

Not all blacks left the Delta or the South. In fact, one even stayed on the farm and kept bees, although unfortunately we don't know his name. A 1959 *American Bee Journal* article mentions a very successful black Virginia beekeeper "who [kept] over three hundred colonies. In fact, he is the largest beekeeper in the Southeast and a good one."[36] In Cade, Louisiana, two black beekeepers—Charles Joseph and Frank Decuir—managed Walter Kelley's queen and package bee industry. Although the company had moved to Kentucky in the 1930s, Frank was in charge of the field colonies used to provide bees for the nucs, and Charles managed the queen-rearing and queen-mating operation.[37] Thus, southern blacks maintained an active and important, although barely visible, presence in beekeeping.

The Appalachian scene seems tame by comparison to Chicago, but honey bees were prominent symbols of sustainability in the literature from that region. In a book targeted to the young adult market, *The Beatinest Boy,* Kentucky writer Jesse Stuart gives one of the first fully developed portraits of an elderly woman beekeeper.[38] In this book, David tries to earn money for a Christmas gift for Grandmother Beverly but finds that he has little heart to kill animals for their skins. David decides to cut down a bee tree and sell the honey. Without realizing David's ulterior motive, Grandmother Beverly offers her help. She teaches him how to cut the tree, take the honey, and repair the cavity in the tree so that the bees will live in it through the winter. She merely passes on the skills handed down from earlier generations, but this story's importance in the 1950s illustrates that honey hunting and hunters were still important in the barter-exchange economy in the Appalachians.

Chicago and Kentucky—the two regions could not be more different. However, in a time known for conformity, suburbs, and social rules, the artists from two regions used the honey bee as a way to express independence, self-sufficiency, and sexuality.

Even though Karl Showler stated that American beekeeping "pretty much collapsed in 1955," international forces were at work that would both strengthen and challenge the beekeeping community.[39]

The 1960s

Within ten years, the order and conformity characterizing the 1950s American Dream (a house, a car, two children, and a wife in the kitchen) exploded just as quickly as it had been created by advertisements, television, and government-sponsored incentives to veterans. Suburbs and televisions created a conformist mind-set that many young people wanted to change when John F. Kennedy became president. Women returned to college to explore new roles and careers. Rock'n'roll, so intimately tied with the black arts movement, introduced the riffs and rills of the southern blues, albeit with electric guitars. International groups such as the Peace Corps sent many beekeepers to foreign countries. These factors, combined with the environmental movement, prompted a cultural shift in values. In the fragments of the fallout, there too were honey bees. The honey bee represented a shift against the industrial countryside that California and, arguably, the United States had become.

To D. C. Jarvis belongs the credit for tapping into a widespread desire for simpler foods, simpler medicines, and simpler lifestyles. When *Folk Medicine* was released, little did Jarvis know how popular his book would become. *Folk Medicine* was a best-seller and prompted a wave of appreciation for bees and honey. Many beekeepers credit the higher honey prices at this time to this book. Although the pharmaceutical industry had long depended on honey as a vehicle in medicines, Jarvis wrote in a folksy tone that was perfect for his audience. Honey was a staple ingredient in many cough syrups. In tonics containing iron, honey could improve the taste. Furthermore, many medicinal liquids could be shaken and their medicinal properties would remain intact if they contained honey. In other words, these liquids had a longer shelf life because honey would prevent microbiological organisms from forming.

Many young people, now called "hippies," returned to farms and cultivated agrarian skills such as beekeeping. Honey became cool. The side effects ranged widely. The number of hobbyist beekeepers skyrocketed. Dick Bonney, for example, left the corporate world to raise bees. Eventually, he became the University of Massachusetts Extension apiculturist.

The back-to-nature movement fueled *Bee Culture*'s staff writer Charles Mraz to develop his own research in apitherapy with bee venom therapy. Even though the medical community did not embrace these holistic treatments at the time, many people sought Mraz for stings to treat arthritis and other ailments.

Beekeepers went abroad as part of the Peace Corps or the Heifer Project. Both programs taught beekeeping in third world countries because there was little cost in investment, low impact on the land, and the rewards were many. American beekeepers also gained invaluable knowledge about politics. Many would learn that their knowledge about bees would do little good in any environment unless people had access to bee veils, gloves, and smokers. When evaluating the impact of Peace Corps beekeepers, *Bee Culture* writer Karl Showler thought that the American beekeepers would get experience in third world countries that would serve them in their own country.[40]

Meanwhile, back in the United States, commercial beekeepers realized that the hippie movement could actually be a good thing, although the effects weren't felt in the 1960s, but rather in the 1970s. Even though honey prices jumped up to fifty cents a pound, as far as the Commodity Credit Corporation was concerned, not much had changed regarding the beekeeper's fate in America. In Whynott's *Following the Bloom,* Horace Bell explains, "For a couple of years, everything stayed the same, queens, foundation, wood. Everything didn't catch up right off."[41]

Meanwhile, beekeepers were losing bees in major kills from pesticides. When farmers were no longer allowed to use the chemical DDT on their crops, they switched to other pesticides. The most dangerous was Penncap-M. This spray, comprised of tiny pesticide capsules the size of pollen grains, was collected by bees and returned to the hive. Exces-

6.1. Medina Honey Fest float (1959). Courtesy of *Bee Culture*. This float's theme, "Free as a Bee," is emphasized by the women wearing shorts and bobbed haircuts. Bees suspended mid-foraging flight are a nice touch. The A. I. Root candle factory is behind the float.

sive bee kills during the latter part of the 1960s were responsible for many commercial beekeepers going out of business.[42] The top four states receiving federal money from the bee-kill indemnity program were Washington, California, Arizona, and Georgia. In fact, bees in Arizona declined almost 50 percent between 1963 and 1977.[43]

Beekeepers found a ready honey market in Japan, another reason why honey prices increased. With a steady export market and a healthy domestic market fueled by alternative lifestyles, beekeepers enjoyed steady prices throughout the 1960s. With forklifts, large trucks, and palletized hives, the migratory beekeeper was the reverse of the back-to-the-land beekeeper. Douglas Whynott explains, "But in the midsixties beekeepers in North Dakota began experimenting with the Bobcat forklifts that had been developed for farm use, machines with sharp turning radii and tilting masts. They developed a hardwood pallet that could hold four to six hives, and lifted the entire unit on the truck. It worked wonderfully, and the beekeeping industry changed in a way it had not changed since Langstroth discovered bee space."[44]

Many women were no longer content to remain inside the 1950s home, however. They wanted to forge new careers and fight for equal rights. Ida Kelley, who had always been a partner in her marriage to Walter Kelley and who had been a home economics teacher before their marriage, volunteered to teach in the local school lunchroom programs that were required to use the surplus honey from the government storehouses.[45] It is ironic that during a time known for disorder and chaos, many women chose to wear their hair in a fashion suggestive of stability and order: the beehive. This hairstyle—no rollers, no curling irons—showed that long hair could be sleek and glamorous. This cultural phenomenon freed women from Donna Reed. Just look at the difference between the women who rode on the Medina Honey Fest float and the 1947 Mormon Centennial. Compared with the antebellum hoops of the Centennial, the Medina queens are indicative of just how quickly American culture changed in approximately ten years. The Medina women wear shorts that are, well, short, and ditto the hair. These women are hatless and gloveless—and more power to them.

The Beekeeper's Daughter

The poet Sylvia Plath, born to a respected German beekeeper, gave voice to a general dissatisfaction among women by using bees as a metaphor.[46] While her marriage to poet Ted Hughes dissolved, she sought solace in beekeeping and poetry. The resulting masterpiece, *Ariel,* is considered her finest collection of poems. The title, *Ariel,* which alludes to Shakespeare's fairy in *The Tempest,* suggests charms and magic spells. However, in Plath's work, Ariel sings a lament, not a magic spell, in "The Bee Meeting":

> The white hive is snug as a virgin,
> Sealing off her brood cells, her honey, and quietly humming.
>
> . . .
>
> The villagers open the chambers, they are hunting the queen.
> Is she hiding, is she eating honey? She is very clever.
> She is old, old, she must live another year, and she knows it. . . .

Many middle-class white women in 1950s households could relate to Plath's queen, caught in a trap of postwar expectations of domesticity, financial security, and families. In a remarkably prescient stanza in "The Arrival of the Bee Box," Plath's inability to put into language the noise of the hive and her mind come together:

> It is the noise that appalls me most of all,
> The unintelligible syllables.
> It is like a Roman mob,
> Small, taken one by one, but my god, together!

Plath's inability to communicate with the bees, her husband, and her therapist finally lead her to conclude: "I am no source of honey, / So why should they turn on me / Tomorrow I will be sweet God / I will set them free."[47] *Ariel* is prophetic, for Plath did set those unintelligible syllables free. However, her verses continue to be a source of comfort for many people feeling the same restlessness that comes with domesticity.

The children's literature market exploded in the 1960s. Lyndon B. Johnson's Great Society policies established Head Start programs, which in turn legitimized children's literature.[48] Before then, as Eva Maria Metcalf argues, children's literature was in a class of its own: "The belief that the child's naïve, unadulterated, and uncritical view of the world should remain intact informed the idea of a separate world of childhood and children's literature, which . . . barred children's literature from mainstream literature and even more so from the privileged status of high literature, which had raised its barriers by celebrating art for art's sake."[49]

Children's illustrators had access to technology as never before. One of Maurice Sendak's first works was a reissue of Frank Stockton's nineteenth-century fairy tale, *The Bee-Man of Orn.* Sendak used watercolors and eighteenth-century fashions, creating a Bee-Man who looks like a colonial American. This reassuring beekeeper—fat, hairy, and jolly—completely revised the image of the beekeeper. The story emphasizes living off the land, being happy with natural products, and taking care of the bees and one's fellow neighbors.

Stories such as *The Bee-Man of Orn* teach about the values of beekeepers in a creative plotline, but do very little in the way of introducing children to the biological facts of bees. On the other hand, according to contemporary beekeeper Kim Lehman, there has been a good body of children's literature about bees that is factually informative—that is, three parts of the body, three types of bees, two types of dances—but these books display little creativity. Lehman suggests that the twentieth century has encompassed both types of approaches to teaching bees to children, but only toward the end of the century has there been a balance of information and creativity, such as is apparent in Joanna Cole's *The Magic School Bus: Inside a Beehive.*[50] This book is creative, but it also manages to impart basic bee biology in a fun, colorful way. Cole's book represents a major shift in approach from nineteenth-century children's literature, which tended to be moralistic and portray bees as industrious, efficient, and capable of stinging if one made mistakes.

The reality of being a bee inspector during this time period was no fairy tale, however. In order to supplement a high school teacher's salary, Stephen Henderson agreed to be a bee inspector in Michigan. The pay was good, but people were still as protective of their bees as they were during the 1930s and 1940s. At one farm, the bees were considered pets, and whenever Henderson would arrive to check the hives, he was greeted with great clouds of goose droppings being thrown at him. Or brooms.[51]

And then came the day when one of Henderson's friends, Ertel Thompson, called to ask for a favor. It seemed that honey bees had settled inside the ceiling of his porch. Normally, this hadn't been too much of a problem—until honey started dripping on his guests sitting outside.[52] Henderson came prepared with crocks, a hive, and a steady sense of purpose. Not so his wife, Carolyn—or even Ertel Thompson. When Henderson took the first panel down from the ceiling, he found that the bees had built a long vertical comb that ran the length of the porch. He quickly found the queen and placed her in his hive, but the number of bees following her into the hive alarmed Carolyn and Ertel. They fled, leaving Steve to finish the job by himself.

During the 1960s, Americans revolted against the conservative values of the previous decade. The entire culture experienced a paradigm shift: feminism, globalization, civil rights, environmentalism, and political protests. This move meant a fundamental redefinition of the American Dream. Americans were becoming more accepting of cultural, racial, and biological differences, but they didn't do so peacefully or gracefully. The decade posed serious questions for the bee industry and for its culture: The honey bee was important, but in what ways? Should the bee be protected, and if so, to what extent? How much funding should Congress allow? What did the honey bee, an ancient symbol of industry, mean to a country that wanted to move away from industrialization?

The 1970s

New York City, 1978. Garbage strike. Trash piled ten feet high. Conditions are ripe for a disaster. Killer bees arrive on a boat from South America. As the bees begin to forage for food, they fly from garbage stack to garbage stack. Eventually, their nests get crowded. These killer bees then attack thousands of people. In 1978, Irwin Allen's *The Swarm* reflected the public's hysteria and helplessness about the advance of African honey bees. The movie's reviews reflected the beekeeping community's ire with Allen's effort: "Hogwash," said the eloquent Tom Sanford, entomologist at Ohio State University, who described the movie as "another brick in the wall preventing communication between the general public, farmers, and beekeepers."[53] In the *Washington Star,* an editorial read: "The notable cast—the biped cast members—has at least the grace to look frequently embarrassed at uttering the dialogue with which they have been afflicted."[54]

Embarrassed adequately describes U.S. officials, who had become alarmed by the advance of the African bees. Barrier proposals were drawn up to stop the African bee from migrating into North America. The first idea was to create a pesticide barrier at the Panama isthmus. The idea barely had time to be written on paper before the African bees swarmed past the isthmus. The second idea was to place thousands of European stock hives across the Panama, in effect a space is needed to create a genetic barrier in which European bees would interbreed with African honey bees and produce gentler bees. Again, this idea was not feasible given how quickly the African bees were moving and the amount of territory that would have to be considered.

What was finally put in place was an African bee trap line, stretching between the Texas and Mexico border. The purpose was not to stop the honey bee, which many knew to be a futile endeavor from the beginning. Rather, officials wanted to be prepared when the African honey bee arrived. It was thought that quarantines would prevent the rapid spread, but even those in place (such as the one in Texas) have come under debate at the time this book was written. Ironically, by the time the African honey bee reached Texas in 1990, it entered with a whimper, not with a bang.

The killer bee controversy generated so much publicity that it served a variety of social functions, however. When Texas Republicans decided to "kill" Lieutenant Governor Bill Hobby's legislation in 1979, Hobby nicknamed the dissident politicians the "Killer Bees." These Killer Bee senators were determined to thwart Hobby's legislative efforts to allow a separate primary day on the issue of redistricting Texas. If a bill for such a primary were to be voted on by the Senate (controlled by the Democrats), the Killer Bees feared minorities might have a tougher time being elected to office. Elected officials would use new census figures to reconfigure districts that would last for at least ten years. "Ordinarily," according to journalist Robert Heard, "the separate-day primary bill could not have been considered because its supporters lacked the two-thirds vote needed in the Senate to bring up the proposal 'out of order.'"

But Governor Hobby invoked an obscure rule that had not been used since 1961: as long as he posted public notice that he would introduce an out-of-order bill for vote, he could and would expect the senators to vote on a separate-day primary. The Killer Bees cried foul. They decided to prevent a quorum by refusing to show up to the Senate. On Friday, May 18, 1979, ten senators did not show up to the Senate floor to vote on the separate-day primary. The majority stayed only blocks away from the capital, convinced that Hobby would let his bill die a good-natured death. But Hobby would not let his plans die so easily. An old-fashioned, dirt-slinging Texas battle was on. Hobby did what any good Texas politician would do: he called the Texas Rangers. The Texas Rangers, a law enforcement unit lingering from Reconstruction times, still had the authority to bring criminals—and politicians—to justice. Hobby ordered the Rangers to find one rogue senator and bring him to the Senate floor during the three days that the bill remained on the docket.

The Killer Bees had not counted on that. They had not brought changes of clothing, toothbrushes, cigarettes, or the necessary diversions that would help them tolerate each other's company for three days. But the Rangers were searching high and low for them. To venture out was risky. The Rangers stopped cars at interstates; they searched the borders. No luck. So the Killer Bees snuck out from their hideout—hiding in trunks of cars, tucking themselves behind car seats, or going incognito—until the separate-day primary died on the floor because Hobby could not call a quorum. The Killer Bees had won!

Consequently, underrepresented groups in Texas had a better chance of appropriate representation. "It was a victory for representative government," applauded Robert Heard, "making it easier for Texas liberals, blacks, browns, and even Republicans to win election to Congress and the Legislature."[55] This political standoff would serve as a precedent for Texas politics in 2003 when Texas Democrats decided to do the same thing, except they learned from the Killer Bees—the Killer "Ds" booked a hotel floor in Oklahoma.

In reality, beekeepers had more to fear from chalkbrood than the African honey bee, although chalkbrood did not make good headline material. In 1972, chalkbrood first appeared among beehives in California. Although in general fungal diseases do not attack living insects, chalkbrood does, using late larval stage honey bees for their food.[56] The disease covers the larvae with a weblike material that resembles gauze. In fact, at this stage, they resemble "soft" and "fluffy" mummies, according to Morse.[57] If the healthy bees do not carry these mummies out, the mummies will get hard as the fungus reaches the reproduction stage. They then will turn black. Chalkbrood gets its name from its initial appearance in the hive, but if the fungus advances, a mixture of white, black, and white-black mummies can be found in the colony. When this fungal disease first appeared, it was made worse by migratory beekeeping. Chalkbrood spread to the other states quickly, although beekeepers quickly learned to deal with its effects, having been warned ahead of time.

Pollination was the name of the game during the 1970s, both in terms of scientific experiments and migratory beekeeping. Dr. Willard Robinson's research with Red Delicious apples brought renewed attention to the importance of pollination. Although Red Delicious apples were developed in New York in 1872, those trees had not been prolific apple producers. Robinson discovered that the Red Delicious has a peculiar blossom structure that prevents honey bees from pollinating it, as they do other flowers. Most apple varieties will not set fruit unless they receive pollen from a different variety. Pollen from a flower on the same tree, or from an adjacent tree of the same variety, will not grow and fertilize an ovary. In old-fashioned plantings, where twenty-seven large trees were usually planted per acre, every third tree in every third row was a pollenizer. In a Red Delicious orchard, however, farmers could plant crab apples, which are great cross-pollinators. They could place more hives in those fields in order to ensure adequate pollination. In short, Robinson's study prompted renewed appreciation of the complexities of pollination.

Combined with the hippie movement, the 1970s were a time of incredible inflation in sugar prices, and hence honey prices. "The real changes in commercial beekeeping were from 1972 to 1980," explains Horace Bell in *Following the Bloom.* "All through those years people came out of the woodwork. Every Tom, Dick, and Harry with a hive of bees jumped into beekeeping."[58]

Meanwhile, more beekeepers lost more bees due to the pesticide sprays. In 1977, E. C. Honl, who got his start with the CCC, lost a thousand colonies because Penncap-M was sprayed on alfalfa.[59] In 1978, a major bee kill occurred in East Central Wisconsin. Several sweet corn companies were located nearby. This case went all the way to the state supreme court, where it was determined that the beekeepers should have kept better records of how many bees they had lost, documented the days on which the bees died, and recorded the damage with a third, objective party present. Because other farmers had not complained about pesticide damage, the company was reluctant to take the beekeepers seriously. This judgment reinforced the disastrous effects that could occur when companies and farmers became careless in their spraying and did not follow the strict warnings posted on the pesticide label.

Incidentally, when the price of honey rose, the government matched the market price. Rather than finding a lot of new markets to sell to, a few beekeepers sold (that is, dumped) their inferior honey on the CCC, which had to accept the honey as payment for loans. Beekeepers who were smart sold their superior grades above the market prices in their own local markets. Not all did this. But enough did so that, in a characteristic understatement, Whynott states, "It was hard for people to accept the idea of a beekeeper with a gross income over a million dollars, and nearly impossible to accept the idea of that beekeeper turning his entire crop over to the government."[60]

There were bright spots in education, however. Women researchers were entering the field, and their research would prove to be beneficial in the 1980s, 1990s, and 2000s. As a student at Preston College in Arizona, Marla Spivak picked up a book about beekeeping, having never thought about bees before. She was fascinated by the bees' interaction as well as the social relationships among the beekeepers themselves. She used a flexible class schedule to allow her to work with a commercial beekeeping outfit in New Mexico for college credit. Sue Cobey, who had a commercial beekeeping background, took a couple of beekeeping classes with Harry Laidlaw and was influenced by his passion for the topic. Both women credit their male colleagues as being very supportive of their efforts to learn artificial insemination, to study African bees, or, later, to support their work on hygienic behavior or disease-resistant bees.

The Vietnam War, like the Korean War, did not affect American beekeepers in ways as immediate as World Wars I and II had. There were no rations, no urgent need for beeswax, and no concern about pollination. However, the effects were to be found in a general recession that affected the availability of bee supplies. In 1974, with the country in recession, "The Vietnam War was taking its toll. Galvanized steel could not be secured. . . . Lumber became difficult to obtain," writes Mary Kay Franklin.[61]

It's no surprise that the roll call of beekeepers is as impressive as in previous wars. After enlisting as a noncommissioned officer in 1958, Stanley Hummer served as a helicopter rescue medic first, and then served as a navigator for the air force. After twenty-two years' worth of missions across the world, "he retired and opened Hummer and Son Honey Farm" in Louisiana.[62] Or at least that is what the obituary states. His son Billy tells another story: When Billy was in fifth grade, he wanted to raise bees as a 4-H project. He and his dad ordered their first colony from Sears and Roebuck. An hour after they arrived, the bees left. Stanley and Billy ordered more bees, and these stayed. Billy credits the military with instilling in his father a respect for order and the processes by which things get done. "As a major," Billy Hummer states, "my father was used to managing people, so it was a good fit to begin managing 'little people'" (that is, bees).[63]

Upon his return from Vietnam in 1970, Larry Walker took over operating his father's, Chase Walker's, business, the Los Angeles Honey Company.

In 1972, a veteran named James Rodden returned to the Odessa Plains with one obsession: to make mead. Although his efforts ended in failure, he spent hours working on his hive. In this way, his friend Chuck Anderson figured, Russ was able to piece a world back together in his hive.[64]

Beekeeper Charlotte Stevenson's fiancé, Vietnam veteran John, should not be left out of this discussion. This is a man who had flown helicopters during Vietnam and who had survived bullets and hand grenades. He felt confident he could handle bees. In a series of very bad decisions detailed in her article, "How *Not* to Move Bees," Charlotte and John let both of their smokers go out in the damp Florida terrain. When bees got inside Charlotte's veil, she fled, leaving John to move the bees himself: "Dark. Angry Bees. No smokers. When [Charlotte] returned . . . [she] found John pressing on, as in Vietnam. And, as in Vietnam, he was losing." Although he was not armed with a smoker, he was determined to move the bees—but not just any bees. Insanely angry bees. John probably does qualify for a Purple Heart. He "steadfastly unloaded those bees amid the fray and paid no attention as [Charlotte] helped then ran, helped then retreated. He never retreated. He took more hits. But the job was done."[65] This narrative has a happy ending: he married Charlotte anyway.

For those veterans who came back from Vietnam, writer Chuck Anderson states, honey bees often provided a way to reconnect with another form of society than the ones in which they had participated in so violently.

Somewhere between the slim, strutting Jimmy Page look-alikes and the Tina Turner wannabes, Van Morrison, the Belfast Cowboy, carved a place for himself in the American rock 'n' roll music scene. Although he was from Ireland, Van Morrison charmed his way into the American mind-set with the sublime "Tupelo Honey" in 1971. Morrison freely admits that his influences were from the American South, citing Leadbelly and the Carter Family (particularly Sara) as his primary influences when growing up in Ireland: "I had this book, it was called *The Alan Lomax Folk Guitar Book*, and it was mainly based on the Carter Family style, which was a picking style, and that's fundamentally the folk style as we know it today. So that's what I was learning. I listened to records as well, of the Carter Family and Lead Belly [*sic*], while I was practicing." Huddie Ledbetter, better known as Leadbelly, had been serving time in Angola Prison when collector John Lomax passed by, hoping to "capture the songs of Negroes with the least contact with jazz, the radio and with the white man," according to Clinton Heylin.[66] Leadbelly fit the bill, and his songs along with the gospel sound of the Carter Family happened to make it into a songbook bought by Van Morrison's father while in Detroit.

So when he moved to the United States in the 1960s, Van Morrison merged the blues and the sensual honey bee imagery with his own Ulster background in "Tupelo Honey."[67] The result is an intense, carefully measured love song that moves as slowly as the liquid itself. Flutist Bruce Royston lulls us into the song. The saxophone solo evokes the sultry swamp atmosphere associated with the Delta juke joints. Although this album was produced during a difficult time in Morrison's life, the song became a Morrison standard. What is important is what the song is not: it is not strictly blues, and it is not strictly Celtic. It is a bluesy little ditty infused with Celtic flute merging into a powerful saxophone-driven resolution. Its sincerity has its roots in the blues, but it is assuredly Irish as well and makes for a nice foundation of the 1970s music, so much of which was blues borrowed unapologetically by rock 'n' rollers.

Morrison wasn't the only one going back to the agrarian images in his music. A group of black women from Washington, D.C., led by Bernice Johnson Reagon, formed their own a cappella group, Sweet Honey in the Rock. Reagon, the daughter of a Georgia Baptist preacher, was involved in the Student Nonviolent Coordinating Committee Freedom Singers in the 1960s civil rights movement. Extending the momentum of the movement to the inner city, Reagon uses all varieties of music—spirituals, blues, jazz, gospels, and hymns—to reaffirm justice and freedom. These women maintained the spiritual and gospel tradition but also acknowledged the changing terrain of gospel: it was no longer just a rural art form, but also an urban one. These women were the best proof of the old parable describing the land of Canaan as being so good that honey flowed when the land cracked open. They were and continue to be strong voices of protest, but their music is smooth and soothing, like a balm for the discord they sing of.

The 1970s began optimistically, with beekeepers seeing profits from natural foods markets and the export market to Japan. But the banter about African honey bees continued to create pressure between beekeepers and the public. Commercial pesticides continued to kill bees, and chalkbrood proved to be a new threat. Beekeepers struggled not only to make ends meet, but to save their bees. Researchers prepared to begin artificial insemination on a widespread scale to address a number of challenges to the honey bees in the coming decades. But toward the end of the decade, beekeepers didn't have political support. When former Wisconsin governor Lee Dreyfus was approached by a well-meaning honey producer, who tried to give him a jar of honey, Dreyfus refused, claiming that his wife Joyce had refused to eat "bee poop" since childhood.[68] Dreyfus survived the furor, but the Wisconsin beekeepers were not happy. Even Georgia-born president Jimmy Carter voted against tariffs that would protect beekeepers from honey imports in 1976.

In retrospect, the 1970s served as a warm-up for the major disasters that would come in the 1980s and 1990s when varroa and tracheal mites arrived. Beekeepers would find themselves very much alone, without political protection, media protection, or even legal protection.

The 1980s

Forget the images of African honey bees swarming helpless bystanders. The tracheal and varroa mites were far worse than anything Irwin Allen could have imagined. However, the process of pesticide regulation got a significant boost in 1981. In a case similar to the Wisconsin case in the 1970s, Connecticut beekeepers suspected that their bees had been contaminated with Penncap-M, the same chemical that destroyed Honl's colonies. The legal process took three years, but eventually beekeepers realized that colony losses had to be documented. The Connecticut group eventually prevailed and the use of Penncap-M was curtailed.

Still, by the 1980s, pollination studies were providing a way to understand how beekeepers could reduce their bee kills. If it is windy, the chemicals can be blown onto target plants. The bottom line: beekeepers had to learn to work with farmers to prevent these kills from happening. But it was often easier said than done.

Congress also revised the honey price stabilization program, which was originally designed to protect pollination services. But as the price stabilization program was utilized during the 1960s and 1970s, the emphasis on pollination evaporated, and honey became its focal point. According to *Bee Culture* writer Glenn Gibson, when the federal program was put in place, Congress did not understand the problems of the beekeeping industry. Many politicians assumed, for instance, that high prices under the honey loan program resulted in overproduction of honey, which wasn't the case.

Letting politicians use high honey prices as the deciding factor as to the availability of honey bees for pollination could be a mistake, Gibson asserted, for honey availability doesn't suggest anything about the number of bees available for pollination. There was so much honey in the 1950s because it was cheaper to let the government buy the honey than sell it on the market, not because bees had produced more honey. However, as a result of unregulated pesticide uses, honey bees were killed frequently. Thus, the hives of bees available for pollination were fewer, although just looking at the abundant amount of honey in the warehouses would wrongly suggest otherwise. Politicians assumed that beekeepers could offset market losses with an increase in pollination fees. Although the idea may be a good one in economic theory, rarely were beekeepers able to raise their pollination prices in practice.

So the government went back to the drawing board. In effect, it wanted out of the honey-producing industry. The Honey Price Support Program of 1986 provided more regulation of the honey industry, especially with loans and buyback policies. If market prices were too low, loans were available for those honey producers who would want to store their honey until the prices were higher. If producers or cooperatives chose to sell their honey, they could repay the loan at below-loan rates.

If borrowers were unwilling or unable to pay back their loan, they could forfeit the honey to the Commodity Credit Corporation (CCC). The CCC was obligated to accept the honey as full payment of the loan.[69] These efforts were designed to provide support to beekeepers while helping the government get out of the honey industry.

Mites

Since 1922, when Congress banned the importation of other varieties of honey bees, the Italian honey bee has been the primary bee used in commercial and hobby operations in America. And although there were studies done in the 1940s to breed foulbrood resistance, chemicals were cheaper and easier to distribute. Very little had been done in the areas of disease-resistant bees until both the tracheal and varroa mites arrived in the 1980s and forced the bee industry to acknowledge the need for such research.

The tracheal mite, which arrived from Europe in 1984, affects the honey bee's tracheal passages, making it difficult for the bee to breathe. The life cycle of the mite works in this way: An adult female mite feeds on the blood of an adult bee's tracheal tube. The mite lays a male egg and female eggs, and then dies. The male and females will mate inside the tracheal tube. The male dies, and the pregnant females continue to feed on bee blood. These mites thus cause damage to adult bees by impairing performance and piercing tracheal tubes.

Unlike the tracheal mite, which affects adult bees, the varroa mite works from within the brood chamber. A pregnant female mite enters the hive on an adult bee. Once in the hive, the varroa mite will bury itself in the bottom of a brood cell, placing itself in the royal jelly. When the cell is capped, the varroa mite will feed on the larvae. It lays several female eggs and one male egg. These will hatch and mate. When the young bee emerges, the mother and daughter mites are ready to begin the cycle again. However, the bee has been damaged because it has been used as a host for the mite.

Of the two mites, the varroa has had more disastrous consequences to honey bees. The varroa mite first appeared in the United States in 1987 from Asia and caused horrific damage, quickly wiping out thousands of colonies. The mite reproduces on worker bees, thus making hives susceptible to mite damage all year long as opposed to seasonally, which tends to happen with tracheal mites. The mites inflicted such damage that many states imposed quarantines, which mandated that beekeepers stay within their confined areas. In North Dakota, the state authorities imposed a quarantine in October 1986. Those beekeepers who were just finishing their summer harvest and preparing to leave for warmer climates were trapped. Some beekeepers lost so many bees to winterkill that the tracheal mite was an afterthought.

Varroa mites inflicted such damage that the Canadian government closed its borders to American queen breeders by 1986. The effect on

American breeders can be summarized in Fred Rossman's experience: "We were shipping fifty thousand packages and queens until the bottom fell out of Canadian exports in April of 1986," explains Fred Rossman, heir of a queen and bee package business. But "when you shut off 90% of your business overnight as they did to us, it leaves you with nowhere to go."[70] Rossman adequately summed up many beekeepers' helplessness when dealing with varroa.

Then there were beekeepers who chose to ignore the quarantines and risk the fines. Driving at night, on backroads, and avoiding public places—these were activities that have a covert ring to them, and in a way, they were covert. Those few who were caught received stiff fines.

Beekeepers and researchers turned their attention to developing resistant bees again. Sue Cobey, having worked with Harry Laidlaw Jr. and John Harbo, polished her skills with artificial insemination to develop a breeding program that would work to select several traits such as disease resistance, good hygienic tendencies, and overwintering. As Cobey explains, normal mating among honey bees can be more complex than people realize because two factors—genetics and environment—make for unpredictable results. In ideal conditions, if a queen bee mates in the air, she would mate with drones that exhibit admirable traits, such as disease resistance, mite resistance, or hygienic behavior. But conditions are rarely ideal, and there are always factors affecting normal mating that cannot be controlled.

It was in the 1980s when beekeepers, especially commercial queen producers, began to take the new stock seriously, and Cobey began to see the potential for developing a good instrumental insemination program for disease-resistant bees.[71] So Cobey turned her attention to a new strain, the New World Carniolans, which has shown some resistance to varroa. They have several advantages: they are winter hardy, take advantage of early spring nectar flows, build up strong colonies quickly, produce good honey crops, and use winter honey sparingly. More has yet to be done with this strain, as varroa continues to be a factor in American beekeeping.

During the 1980s, Marla Spivak attended the University of Kansas, "a mecca for social insects," to use her words. After landing a grant to study how African bees would respond to higher elevations, she went to Costa Rica in 1984. Interestingly enough, in her studies she learned that

African bees adjusted quite well to higher elevations and that the Costa Rican people adjusted quite well to the African bees. In fact, the biggest problem associated with the African bee, it seems, is that Costa Rican people did not have the proper equipment such as bee veils and suits.

Meanwhile, a pest management system used in a variety of other agricultural crops and parasite control was beginning to be adopted by beekeepers. Integrated Pest Management (IPM) was developed in order to minimize chemical use for mite control and promote mite resistance in the hive. Briefly defined by Tom Webster, a Kentucky apiculturist, IPM describes the "optimal combination of methods used to fight a pest or disease."[72] IPM differed from previous approaches to hive management because it didn't rely on a calendar, but rather it promoted using three or four methods simultaneously. Beekeepers were advised to medicate hives only when varroa mites had reached threshold levels. And in lieu of destroying all mites, beekeepers were encouraged to add to their hives' defense by tolerating the mites until the hives began to suffer from stress.

Although the honey bee suffered in the real world, it did not suffer in social and cultural usages. Writers, artists, and photographers championed its image, using it occasionally to suggest thrift in an unthrifty society, but also more often peace, domesticity, and political astuteness.

In 1980, Hal Cannon brushed the dust off the bee skep symbol in Salt Lake City. In order to promote its importance, Cannon (a descendant of Brigham Young) arranged for an exhibit devoted to the artistic use of beehives at the Utah Folk Art Museum. The Grand Beehive exhibit provided three different types of beehive-related shows: nineteenth-century artifacts, artistic works, and newly commissioned pieces created for the exhibit. In summing up his experience with this exhibit, Cannon stated: "I found a few people who felt as strongly about beehives as the [Mormon] pioneers did, who always loved the symbol, had beehives in their homes. . . . It's a sentimental thing, connected to Mormonism, and I became a little sentimental about the beehive myself."[73]

It is impossible not to be. Travel to any museum in Salt Lake City and you can be overwhelmed by the woodwork of Ralph Ramsey, the most famous of the Mormon woodcarvers. A bed made for Brigham Young is replete with beehives on the posts and bottom board. Ramsey's best-known sculpture is the beehive that sits on top of Eagle Gate, the entrance to the Utah state capitol.

6.2. Utah shell game. Courtesy of the Utah State Folklife Archives, Utah Arts Council. Photo from *The Grand Beehive*, Brent Herridge, photographer. David Pendall's sculpture was a political statement, protesting the state's consideration of moving missiles around Utah. It was part of the 1980 Grand Beehive exhibit in Salt Lake City.

Cannon invited contemporary artists to submit original pieces using the skep as a central image. Sculptor David Pendall, using this invitation to comment on Utah's proposals to relocate missiles, created "The Shell Game." In his exhibit, "Three bee skeps, four feet high, with a missile bursting through one of them . . . comment on Utah's resistance to the MX missile deployment in the state."[74] This creation was a response to some Utah state politicians, who had considered moving bombs around Utah in an attempt to fool Soviet defense strategies.

The Grand Beehive exhibit was wildly successful. The Smithsonian featured it in the Renwick traveling display in 1981. But more importantly, according to Folk Art Museum curator Carol Edison, the larger goals of the exhibit had been fulfilled. People were reminded of their heritage and the importance of the image when Utah was still a fledgling state. Edison remarked that artists were starting to use the beehive again, and just to prove her point, she whipped out an article on the best pies in

6.3. Snelgrove's bee hive ice cream. Courtesy of the Utah State Folklife Archives, Utah Arts Council. Photo from *The Grand Beehive,* Brent Herridge, photographer. For fifty years ice cream was served in this shape at Snelgrove's.

the Salt Lake region. In the visual layout of pies in the Utah landscape, the hills were subtly shaped like skeps.[75] Even ice cream could be sculpted into skeps.

Cannon and Edison were not alone in revering bees. In 1985, photographer Richard Avedon decided to capture the New West in an effort to understand how the land and its people have changed since the closing of the open range. "Right from the start, Avedon chose men and women who work at hard uncelebrated jobs, the people who are often ignored or overlooked."[76] There aren't any clean cowboys in this photography collection. Technology and service-oriented jobs dominate this photographer's landscape *In the West.*

So it is somewhat of a surprise to come upon the picture of a beekeeper in the midst of these working-class people in the New West. The portrait "Buddhist" redefines the bee-hunter stereotype in innovative ways. First, Avedon chose a beekeeper named Ronald Fischer who is as bald as the old nineteenth-century bee hunter was supposed to be hairy. Next, Avedon photographed Fischer covered in bees against a white backdrop. The portrait shows a white man, completely without fear, so completely at ease with his bees that one wonders where the bees end and the body begins.[77]

The *In the West* photographs, first exhibited and published in 1985, generated controversy. Avedon said, "I don't think the West of these portraits is any more conclusive than the West of John Wayne." Some critics praised the portraits' rugged individualism and pathos, but others hated the exhibit. "This is a sick collection that expresses Avedon's inner fears and terrifying nightmares," fumed Fred McDarrah, the original photography editor of the *Village Voice.*

If indeed McDarrah is correct, Avedon's inner fears seem to be absent in "Buddhist." The stereotypical bee hunter of the Wild West conjured images of shaggy, bearded men living on the margins of society. But Avedon offers in his photograph the complete opposite: an unrobed man covered in bees. This image of peace, solitude, and a momentary oneness with nature complements the other photographs of raw, gritty, bare-knuckled people struggling to be independent souls. In the New West, only the beekeeper promises an affirmative way to blend the two ideals of work and independence, Avedon suggests.

Similarly, with her spare writing style, Sue Hubbell created a lens through which to view the Ozarks in *A Country Year.* Hubbell, a female beekeeper in a predominantly male field, recorded many social changes experienced by women beginning in the 1970s. Domestic entanglements, gender differences, an eventual divorce, an eventual new beginning—through all these troubles, her bees stayed the course. With the success of *A Country Year,* Sue Hubbell began writing a book strictly about beekeeping. *A Book of Bees* was published in 1988, and at first glance, it seems remarkably similar to Mary Louis Coleman's *Bees in the Garden and Honey in the Larder* (1939). Both are arranged according to the seasons, and both have intelligent, unpretentious writing styles. Despite the differences in geography, both women share the same frustrations and pleasures with beekeeping. The differences between the two are just as apparent, however.

An Ozark beekeeper, Hubbell is not worried about the three Cs (cooking, cosmetics, and cleaning). In fact, she is a divorced woman working her way toward a new life by writing and taking care of bees. In *A Book of Bees,* Hubbell provides a nice balance of history, literature, and practical instruction so that anyone could learn more than how to manage a beehive. The Ozarks are an important context in which this narrative is written:

> The store in the little town between two ranches carries more than groceries. In the back of the shelves of peanut butter and canned corn are work clothes, axe handles, coils of rope and other country necessities. In the middle of it all is a Formica-topped table, where ranch hands are usually gathered drinking coffee from the pot that sits on a hot plate at the end of the meat counter. . . . I don't know their names, but they know mine: Bee Lady. A middle-aged woman in baggy white coveralls who smells of burnt baling twine is a standout in any crowd.

At first, Hubbell stands out in the Ozarks, but she finds that she fits in quite well. She can talk weather with the ranch hands, and that is the real key to establishing friendships: "So we talk weather. We talk hay. We talk bees. We talk farm prices and shake our heads sadly."[78]

Hubbell's book begins in autumn, but the timelessness about the art of beekeeping comes through in the writing, regardless of the season. No matter when a person begins beekeeping, Hubbell suggests, there are challenges and joys. In the "Winter" section, one learns that bees are "forgiving" creatures, and will tolerate a good deal of gadgets, but the "best beekeepers I know are those who let the bees themselves, not equipment manufacturers be their teachers." In fact, she says in understated Ozark humor, "The only time I ever believed that I knew all there was to know about beekeeping was the first year that I was keeping them."[79]

Hubbell's book is a practical guide to beekeeping, but it is also a love story. She loves her bees, her community, and the Ozarks. In the end, her love is returned. She hears from an old sweetheart, and *A Book of Bees* ends, appropriately, with discussion of nuptial plans of the queen bee.

Similarly, Fannie Flagg's *Fried Green Tomatoes at the Whistle Stop Cafe* features an androgynous female bee charmer. The novel begins with Evelyn who, having been married to the same man for twenty years, has lost her zest for life. She wanders over to the Rose Terrace Nursing Home, where Mrs. Threadgoode's stories about Idgie and Ruth mesmerize her.

Idgie and Ruth make up the second plot line, set in the early 1940s. These two women break every social rule determining feminine behavior in the rural South. A tomboy, Idgie, the bee charmer, works her best magic in the woods: "She very slowly tiptoed up to [the tree], humming very softly, and stuck her hand with the jar in it, right in the hole in the middle of the oak. In seconds, Idgie was covered from head to foot with thousands of bees. . . . By the time she had gotten back, almost all the bees had flown away and what had been a completely black figure was now Idgie, standing there, grinning from ear to ear with a jar of wild honey."[80] Idgie's offering of wild honey shows her complete, unconditional love for Ruth. When Ruth is clearly enchanted by her friend's courage, Idgie was "as happy as anybody who is in love in the summertime can be." Idgie's crush on her friend is not allowed to develop given the pressures in a small southern town. But when Ruth marries a ne'er-do-well who beats her, Idgie hatches a plot to save Ruth and open the famous Whistle Stop Cafe.

The bees play a minor role in this novel, but the movie had a powerful effect on women during the 1990s. With such stars as Jessica Tandy and Kathy Bates paired with relative newcomers Mary Stuart Masterson and Mary-Louise Parker, many marginalized feminine topics were in the spotlight: lesbianism, old age, segregation in religion, and obesity. Most important, however, is friendship.

Evelyn overcomes her own fears after hearing about Idgie's hopes and disappointments. Finding herself drawn to a black worship service, Evelyn "felt joy. Real joy. It had been joy that she had seen in Mrs. Threadgoode's eyes, but she hadn't recognized it at the time. She knew she might never feel it again. But she had felt it once, and now she would never forget the sensation as long as she lived."[81] From this point forward, Evelyn takes control of her marriage, her body, and her attitude. But the scene that keeps this movie from being a stereotypical chick flick is Idgie's role as a bee charmer. Her courage, her natural assimilation into nature—Idgie is capable of finding joy amidst the hardships of life. The scene that best pays tribute to her skill is the last one, in which a jar of honey is left by the dilapidated remains of the Whistle Stop Cafe.

Appalachian writer Sidney Saylor Farr adds her distinctive voice in *More than Moonshine,* which focuses on the Appalachian diet that she grew up on in the 1950s. In a book very similar to Jesse Stuart's *The Beatinest Boy,* Farr recalls how people were able to determine ownership of wild bees during the early part of the twentieth century: "When any of the Saylor men found a bee tree they cut two parallel vertical slashes in the bark. This mark let anyone else . . . know that the Saylors had claimed this tree.[82]

A Kentuckian by birth, Jesse Stuart provides a remarkably different twist to the old bee hunting tales of the nineteenth century in a coming-of-age story, "Saving the Bees." Three teenagers steal people's bee gums or saw down bee trees that have been marked. "Saving the bees," according to Big Aaron and his buddies, means releasing the bees: "It is a shame to coop bees in boxes and sawed-off logs and make them work their lives away for a lazy bunch of people. We won't have it!"[83]

One by one, each farmer is singled out. The boys "save" fifty beehives from local farmers and set the hives out in the forest. The typical male posturing happens between the teenagers as they determine manliness by how heavy the bee gums are, how much courage they, as bee thieves, have, or how much ingenuity they have when parents and neighbors question them. The tiny Appalachian community is in uproar: an unwritten social code that protects bee trees has been violated. When one man says he doesn't know what the world is coming to, his wife answers, "It ain't comin' . . . it's goin' . . . and to the Devil it's goin' fast. We ain't had sicha thing happen for years."[84] A moonshiner takes the blame for stealing the beehives, and the teenagers are let off the hook.

In Robert Morgan's poem, "Moving the Bees," the Old World funeral tradition once again finds a place in the twentieth century. When the beekeeper in the poem dies, the protagonist goes to shift the hives. Death in this poem is not the gussied-up ceremony that funeral parlors have dictated that it should be. Rather, the beekeeper dies with a simple farewell, suggesting that the process of dying is nothing more than another small movement. The protagonist in the poem explains:

In January cold
or March wind or summer dark the gums
must be shifted an inch, a finger
width, from the place the keeper left,
them or the colonies would die
or swarm and leave.[85]

The bees are so sensitive that this movement registers the death, along with the passage of the seasons, the sun, and the dances. Far from a lonely passing, the old beekeeper takes one last breath, "and the universe was moved." Morgan plays with two different meanings of the word *moved* (the motion and the emotion) in the concluding line. But I think on a broader scale Morgan moved his audience back to a simpler time when death could be marked by a simple goodbye.

While the gospel group Sweet Honey in the Rock continued to become more popular in the 1980s, blues took a back seat to rock 'n' roll, except for a skinny Texas dude named Stevie Ray Vaughan. Vaughan, who grew up with blues legend Albert King as an influence, barreled out a primal version of "Honey Bee," which had a lot in common with the blues players of old. Although the blues had been marginalized as a genre during the 1980s in favor of synthesized sounds, Vaughan bucked the tradition, going back to the southern black guitarists and the rural images they were surrounded by. In this peppy song, the protagonist in the song begs his "queen bee" for a kiss, "cause we just can't miss." Claiming that her gift is the ability to give him a buzz, Vaughan strolls off with a rollicking guitar solo, in effect creating that buzzing sound on his six-string. Rather than end the song with a tune going down the scale, Vaughan delicately plays up the scale, leading the song and the listener into an imaginative ending, where it is hoped the queen and her adoring drone fly off somewhere in the Texas cattle cradle. This song then ends like many blues songs in which, as Vaughan himself once described, "music used to be based on common everyday occurrences like the train sound going down the tracks, someone walking down the streets, a horse walking . . . that's where these rhythms came from."[86]

The beekeeper's "blues" began in earnest with the arrival of the tracheal and varroa mites in the 1980s. However, in terms of cultural shifts, the nation reverted back to the honey bee, using it as a way to help people connect with nature, the past, and themselves. Perhaps no one did the honey bee better than Roxanne Quimby, who, with clever packaging and solid business principles, started Burt's Bees in Maine in 1989. After approaching Burt Shavitz for beekeeping lessons, Quimby decided to apply her artistic talents and business skills to start her own cosmetics line. The result was a line of high-quality, bee-related products, often made from old *American Bee Journal* recipes, and a folksy package. "Little did she know that 14 years later," writes Howard Scott, "she would be presiding over a $55 million dollar empire."[87]

The 1990s

The plagues affecting bees in the 1990s created a crisis of epic proportions: varroa mites, tracheal mites, federal quarantines, African honey bees, import woes, and, last but not least, small hive beetle damages. Even El Niño created havoc. As honey bee populations were declining in the United States because of mites, domestic beekeepers were competing with foreign imports from Argentina, China, and Mexico. As if those two pressures were not enough, beekeepers finally had proof that bacteria causing American foulbrood had developed resistance to Terramycin. Integrated pest management, once an option, now became a requirement.

Globalization was the key concept in 1990s honey production. Compared with other agricultural industries, the beekeeping community always has been small. Whereas other farmers could hire cheaper migrant labor or buy newer labor-saving technology, beekeepers simply have not had those options. Beekeepers must train their workers to care for bees, and their skill and knowledge can have immediate implications. Furthermore, pollination has not been an industrial service the way that other agricultural services have become. "Very few contracts will ever get written in paper," explains Eric Mussen. "Pollination is still very much a verbal agreement between an orchard grower and a beekeeper that often has developed over years."[88] So the beekeeping community in the United States became even smaller in the 1990s as older beekeepers retired, bees died in record numbers, and competition with other countries increased.

Furthermore, within this small community, sharp differences of opinion about how to approach problems of importing honey came to the surface in the early 1990s. Although domestic honey production was down because of the mites, the weather, and fewer beekeepers, many domestic beekeepers were upset with cheaper honey being imported into the country. Although there had always been splits in the beekeeping community, the question of honey tariffs solidified two distinct groups: the American Honey Producers Association and the American Beekeeping Federation.

The American Honey Producers Association represents the interests of domestic commercial beekeepers and backs price supports and tariffs. Yet this group could not possibly provide the honey needs of the country. It also could not serve the niche market created by raw honeys or varietal honeys. Yet its members wanted (and would still like) more oversight of honey being imported from China and Argentina.

The American Beekeeping Federation represents much of the industrial interests of beekeeping: the honey packers and importers, the queen and package producers, and a large number of commercial and sideline beekeepers. Although tariffs were proposed to keep imported honey out of the United States in order to support domestic beekeepers, the American Beekeeping Federation also recognized that some honey would have to be imported to meet the country's needs.

As a government-sponsored marketing group, the National Honey Board (NHB) cannot lobby on the issue of tariffs. Instead, it promotes the use of honey through generic advertising and data for honey users, commercial honey producers, importers, packers, and the public. It also sponsors research projects to improve the use or increase the use of honey. The NHB sponsors round tables for the American Honey Producers Association and the American Beekeeping Federation, but it was not able to fund the legal research necessary to convince Congress to pass honey tariffs in the 1990s.

Migratory beekeeping continued to be the most profitable way for American beekeepers to make money in the 1990s. The wanderlust of migratory beekeeping was effectively captured in *National Geographic*. Writer Alan Mairson interviewed Joe Tweedy and Jeff Anderson. These men and their families used eighteen-wheelers loaded with a thousand hives as they migrated from California to Minnesota, following the crops and providing an "essential but unheralded" service to American agriculture.[89]

Before the almond industry, Mairson reports, farmers would rent their hives for $2 a hive. But because of new research showing how important pollination was to orchard growers, prices climbed. During the 1980s, rental fees climbed to $20 a hive during the lucrative almond season. In the 1990s, Tweedy and Anderson charged $32. With all the beekeepers combined in California during February and March, "The almond pollination bill could exceed 25 million dollars," Mairson sums up.

Mairson's article was popular for many reasons: good, clear charts explained how migratory beekeeping worked; photographs with stormy clouds and golden hives created attractive visuals; statistics showed how little money beekeepers earned; and perhaps above all, the article emphasized the beekeeping family and its passion for bees. Even though Joe Tweedy had been in the business for thirty years, for example, he and his wife did not have huge savings accounts in the 1990s. Although migratory beekeeping has changed since this article was written, Mairson's emphasis on family remains true. Many commercial beekeepers were and remain family operations. If you want to save family farms, commercial beekeeper Richard Adee cautioned, save beekeepers, whose businesses are primarily family oriented.

After forty years of publicity and fanfare, the African bee made its way to Hidalgo, Texas, on October 15, 1990. Two members of the African Honey Bee Information Team—Evan Sugden and Kristina Williams—documented the day the African bee arrived, which started out like any other routine day of checking African honey bee detection trap lines.

"The sixteenth stop was made at a somewhat dismal spot, distinguished only by two trees on the south side of the road and an irrigation canal bending to the north. Recently disked cotton fields and mature sugarcane dominated the horizon, except for a distant band of dark green foliage marking the twisted course of the Rio Grande. One bait hive unit hung in a small mesquite, the other in a gangly *huisache,* or acacia."[90]

Two field technicians intentionally disturbed these bait hives, waving black suede-covered patches in front of the bees: "The two patches attract guard bees like so many iron filings impacting a powerful magnet."[91] The patches were later found to contain over a hundred stings. When these bees were confirmed to be African, scientists set in motion an entire sequence of information, quarantines, and media relations campaigns.

Comparing the African bee to the Italian and German bees first brought to America in the 1600s, bee historian Gene Kritsky suggests that the two movements were similar to each other in the way they move along the coasts: "In an interesting parallel," Kritsky notes, "the initial spread of the honey bee in North America was also along the coast." By the time the first colony of African bees had appeared, fortunately, much of the worst publicity had passed because the tracheal and varroa mites had been so destructive. Some states tried to impose quarantines, Texas and Arizona in particular, but Kritsky cautions, "If history teaches us anything, it is that the African honey bee will not be confined by human activities but rather by geographic, climatic, and biological factors."[92]

Furthermore, Marla Spivak cautions that the African bee has been so stereotyped that there has yet to be a lot of research concerned with breeding those African colonies that are not as defensive as popularized by the media. "There is a lot of variation among colonies," she notes.[93]

The efforts of earlier twentieth-century researchers were not in vain. Many researchers went back to the 1930s and 1940s genetic studies in disease resistance to begin to find ways of dealing with mite and chemical resistance. In 1997, for instance, several honey bee colonies with American foulbrood in Wisconsin did not respond to the chemical cure Terramycin, which had been so effective in curing the bacteria since the 1950s. The first reported cases were in small bee operations in the northern Midwestern states. But resistance was soon found in other large migratory beekeeping operations shortly thereafter. Resistance to Terramycin, then, encouraged small-time beekeepers to implement integrated pest management plans in an effort to contain the problems of pests and diseases.

Entomologists and researchers branched out in different ways to try to find and breed new strains of honey bees in which good traits, such as disease resistance, ability to winter well, or gentle temperaments, could be defensive mechanisms. In that way, beekeepers could get away from using chemicals. Beginning in the 1990s, U.S. Department of Agriculture (USDA) researchers John Harbo and Jeffrey Harris have been working on a trait called suppressed mite reproduction, which stops the mite from reproducing. "Finding natural resistance has been our goal since the mite moved into the global population of honey bees," explained

Harbo.[94] However, because honey bees were not showing resistance, Harbo and Harris tried to find factors that bees use to control the mite damage. One of those was nonreproducing mites, in which mites enter the brood cells to reproduce, but do not. Harbo says it remains a mystery why the mites don't lay eggs in these honey bee colonies, but he figured there was a genetic component within the honey bees that prevented, or rather suppressed, such reproductive action taking place in their cells.

A technique among bee breeders is called single-drone insemination. Erin Peabody explains: "Using a special instrument, the scientists inseminate a queen with the genetically identical sperm of one drone. Her progeny are then very closely related and are more likely to express genetic uniformity at the colony level."[95] So when Harbo selected drones from colonies that showed resistance to varroa and mated them with a queen who was also resistant, he increased his chances of having colonies that would show resistance once the queen would lay her eggs. His expectation at the beginning of the tests in 1995 was that there would be incremental resistance to varroa mites. However, he was surprised at the "almost total resistance" he was able to detect after a couple of tests. "With single-drone insemination, 1 in 10 colonies had very low levels of mite reproduction. With natural mating, I estimate that this level of resistance would be seen in 1 of 1,000 colonies."[96] Harbo's study is ongoing.

Another development in genetic breeding was hygienic behavior. Although most honey bees can detect diseases within their colonies, Minnesota researcher Marla Spivak distinguishes between two types of bees: hygienic bees and nonhygienic bees. Nonhygienic bees tend to be the honey bees that react very slowly to a disease such as chalkbrood. The nonhygienic bees will wait until the chalkbrood becomes infectious and only then attempt to carry the infected material out of the hive. However, by this time, the disease is infectious and the bees inadvertently spread the disease throughout the hive.

Hygienic bees, on the other hand, detect diseased larvae before they become infectious, and quickly remove the affected area before the hive suffers. Good hygiene is a trait that can be passed on genetically. By selecting drones from hives that show good hygienic traits, Spivak artificially inseminates queens and then provides these queens (carrying the drone semen) to commercial queen producers. The commercial beekeepers can then raise addtional queens from these "breeder queens." These queens will raise worker bees with this hygenic behavior.

The most recent threat to honey bees was in 1998 when four states were confirmed with the small hive beetle—Florida, Georgia, North Carolina, and South Carolina. Small hive beetles are a different kind of threat because it is extremely adaptive. An adult beetle enters a hive and lays eggs. When the eggs hatch, the larvae can destroy a hive by tunneling through wax comb, eating anything in its path—bees, wax, and bee larvae. As if this destruction were not enough, as the beetle larvae tunnels through the hive it will also leave behind a lubricant that adds moisture to the inside of the hive and creates conditions for a "violent ooze," to use Kim Flottum's words. If the hive is not destroyed, the beetle can completely demoralize the hive so that the bees will abscond. The beetle larvae, when mature, can leave the hive and pupate in the soil, or it can live for a long time in the hive without being prompted to lay eggs. However, when the hive is threatened by something stressful (heat, drought, mites, an inspection, or even a harvest), the beetle will lay eggs. If necessary, the beetles can form a protoswarm and take flight to look for another hive to take residence in.

These threats—small hive beetle, trachael and varroa mites, and African honey bees—have forced many people to reconsider the Honey Bee Restriction Act of 1922. In *Waiting for Aphrodite,* Sue Hubbell interviews a scientist who still struggles with the consequences of the Honey Bee Restriction Act. Scientist Suzanne Batra would like to import pollen bees to help address the pollination need in the wake of the honey bee devastation caused by these invaders, but she cannot. She makes the link between the 1922 restrictions and social conditions: "Some people have a prejudice against exotics. . . . It is an anti-immigrant feeling. But after all, most of our crops, livestock, and we ourselves are exotics."[97]

Sue Cobey also has great difficulties with the 1922 Honey Bee Restriction Act. She would love to import the *carnica* race, but only the USDA can import different races. "It is unfortunate that in Europe, even small-time hobbyists enjoy specialized lines of honey bees, some with pedigrees going back to 1948," states Cobey. "In America, by contrast, bees are primarily part of a mass industry."[98] In 2004, the USDA recended much of the 1922 Honey Bee Restriction Act.

Fortunately, the bee charmer was alive and well in literature, films, and music during the 1990s. On the music scene, Florida rock 'n' roller Tom Petty extended a tradition started with Slim Harpo. Although Petty grew up an Elvis Presley fan, which influenced his early rockabilly days, he went through a serious blues craze in the early 1990s. On his *Wildflowers* release, the "Honey Bee" is all about forbidden sex in the Promised Land. The protagonist, a king bee, begs his queen: "Give me some sugar, little honey bee":[99]

> Look here now, peace in the valley
> Peace in the valley with my honey bee
> Don't say a word, 'bout what we're doin'
> Don't say nothin' little honey bee

This album represented a shift for Petty, who left his band The Heartbreakers to record *Wildflowers* by himself and marked his return to acoustic blues and rural images. Petty, reaping the benefits of MTV's popularity, was better known in the 1980s for sci-fi videos in which parachute-clad models drove space-age cars. But in *Wildflowers,* and with "Honey Bee" specifically, Petty strips all the space-age façades away and acknowledges Florida, the powerful bee image, and its effect on the Florida landscape in his music.

Who would have guessed that disco queen Gloria Gaynor would quit singing "I Will Survive" long enough to sing her own version of "Honey Bee"? Gaynor, from New Jersey, sings from the perspective of the queen bee, although there doesn't seem to be a power struggle in this song as we've seen in the others. The queen bee affirms that "love is where you are / There's where I want to be." In other words, this song is not so much about forbidden love as it is about mutual love. There is no fear in this song, no bitterness, and no self-righteousness. Gaynor delivers a simple but powerful affirmation of the goodness in mutual attraction in "Honey Bee."[100]

Southern writer Lee Smith used the bee charmer in *Fair and Tender Ladies* to help a woman negotiate a midlife crisis. The novel is a series of letters written by Ivy Rowe, an Appalachian woman growing up in the early twentieth century. During the 1940s, Ivy begins a descent into unhappiness. Outright rebellion is not the name for Rowe's unhappiness, but dissatisfaction with the state of the world comes close. The 1940s bring incredible changes. According to critic Tanya Long Bennett, capitalism changes the region "to a place where work is not to be found and one must look after oneself since the community is not made up of mountain folk anymore but mountain folk mixed with people who have found that the mountain country is a good place to 'make a buck.'"[101]

Ivy Rowe leaves her husband to follow a beekeeper named Honey Breeding. He is everything that the old-time Appalachian men used to be: self-sufficient and solitary, he follows his own time schedule. Honey Breeding, a World War I veteran who suffers from psychological wounds too deep for any traditional institution to heal, provides a good balance to the technological changes taking place in the region. On a mountaintop, Honey and Ivy culminate their attraction for one another, each finding in the other a balm for the anxiety that the new times bring to their old roles. Honey declares Ivy a monarch in the shifting cultural setting: "You will be a Queen," he says, placing a laurel crown on her head. Then Honey feeds her stories the same way that a queen is fed royal jelly. After admitting she's starved for such tenderness, Ivy says, "It seemed like years since I had heard a good story."[102]

Ivy finally realizes that the difficulties of being a wife and mother leave her exhausted: "I'm tuckered out with lovin," she says finally. "There's so much love in the world, I'm fairly tuckered out with love." But after her sojourn with Honey Breeding, Ivy returns to her husband a new queen. She is a better mother and wife.

Honey Breeding shines in this novel as much as Ivy: he is an Appalachian Aristeaus, the Greek god of beekeeping. Both Aristeaus and Honey grow up without parents but were adopted by families who taught them beekeeping. Their skills afford them special places in the communities. Both balance the social order of the bees with the solitude of the wilderness. Both disappear without a trace.

Because this book began by discussing the importance of Greek mythology and Aristeaus, I think it is appropriate to end with a discussion of the sublime film *Ulee's Gold* (1997). The screenplay is based on the Greek myth of the war hero Ulysses as he makes his way home after twenty years of wandering. In the Greek myth, his wife Penelope has been courted by many suitors in his absence, but she has stayed true to Ulysses. Thus, although the myth is assuredly about heroism and courage, it also emphasizes the importance of home, fidelity, and the difficulties of maintaining those in the everyday world.

Independent Florida filmmaker Victor Nunez carefully creates a parallel situation with a Vietnam vet named Ulee.[103] Having resettled into his beekeeping business in a small town in Florida, Ulee struggles to make his way back into the unfamiliar territory of the home, and ultimately the heart. His granddaughter, Penny, is the constant reminder of why he needs to stay consistent in his values. But he finds himself tested when his daughter-in-law Helen gets in trouble with drug addiction.

Ulee would prefer not to deal with his daughter-in-law. He is in the middle of the tupelo season, the best commercial honey he has to offer. He has decided that if his son and his wife have chosen drugs to cure their problems, he will write them off. When he gets the call from his son, who is in jail, that his wife is in trouble, Ulee hesitates to go pick her up. He does not want Helen, his daughter-in-law, in his home. "Ulee is bad in the sense that he is someone who does not forgive mistakes," explains Peter Fonda, who played the role of Ulee. "He is bad in the way that a Baptist preacher can turn bad."[104]

Thus, although this movie is about beekeeping, therapy, and war veterans, it is also just as assuredly about requeening a broken home. Ulee's beloved wife is dead; his daughter-in-law Helen is a drug addict. Although Ulee would prefer to take care of his family's problems by himself, he finds himself forced to accept the help of his next-door neighbor, Connie, who happens to be a nurse. Patricia Richardson, who played Connie Hope, explains the purpose of the three women: "Ulee is sort of beset by these women in the movie. He is trying to raise his son's daughters, he's got this volatile daughter-in-law to deal with, and this nurse who suddenly comes into his life. These women, in their own ways, force him to cope, to come back to life."[105]

The bees provide the subtle background metaphor for the larger human family. When asked how his bees are doing, Ulee responds: "The mites getting them. The bears getting them. Pesticides getting them. They're doing fine." His answer is perfect for the postmodern family as well, threatened by all sorts of outside dangers. Although it is dangerous to carry metaphors to an extreme, Ulee's answer applies equally well to the contemporary family as it struggles to stay together in the midst of difficult times.

When his family is threatened by two ne'er-do-well criminals intent on evening a past score, Ulee very carefully draws the threat away from the women and manages to trick his opponents in subtle ways in the dark Florida swamps. But the more difficult challenge is not in the swamps but in his heart. The movie's resolution shows how difficult and complex the process of healing and forgiveness can be, but also how very profound, especially when grandchildren are involved.

I too have been a granddaughter working with a beekeeper, finding refuge among his bee colonies. In June 2000, my grandfather's World War II buddies gathered in his honey room. They were old, shoulders gently sloping to the floor, eyes brilliantly bright. They talked in measured sentences of being eastern Kentucky boys, sent to Australia, where they ate shrimp and got into good-natured fights in bars.

While in Guam, these men formed a medical team, and in their company, my grandfather became the man they knew then—relaxed, friendly, profoundly peaceful. They had returned to the States, to their girlfriends, and to the GI Bill, which paid their way through college and taught them progressive farming techniques. But these men were never content to be just farmers; they were teachers first and foremost.

During that summer, they taught me. They talked honey, bees, and pollination. They shuffled to the barn to collect their jars of honey as payment for letting my grandfather keep his bees on their places, but it was a grand occasion nonetheless. He milked every minute of their friendship, showing off his apple trees, his granddaughter, and, of course, the honey we had bottled. He knew it was his last summer, and they did too. They invited him to a World War II veterans' reunion; his eyes burned blue at the thought of it. He did not make it.

For a week after my grandfather had died, I worked the bees alone. I did it mechanically, hearing his slow voice control my actions: "Easy now, open the super gently, a little more smoke." The bees tolerated my awkwardness with the smoker, the hive tools, and the bee brush. One week passed before I told the bees their keeper had gone. I did not drape the mourning clothes over the hives when my grandfather died. I finally learned that Old World European customs associated with bees remain important in this country because the emotional needs have remained the same. Regardless of how technological our society becomes, our participation in this culture will have both sweetness and sorrow. No

6.4. In the key of bee, courtesy of *Bee Culture*. Entomology professor Norm Gary plays for an attentive audience. This picture was taken during a rehearsal for his performance at the Orange County Fair, the theme being "How Sweet It Is."

other insect best expresses the price of being a member of society. No better provides a cure.

The Old World traditions—telling the bees, moving the hives an inch, and leaving jars of honey—still had a place in the twentieth century, for they are about movement within those trying to move on as much as for those who have gone on. With the repetition of the exercise in the dying day, I finally acknowledged the Bee Master would no longer stand beside the hives, nor would he stand beside me.

Throughout the twentieth century, American bee culture underwent radical changes in short bursts. Whereas the first half was associated with industrialization and isolation, the second half was about technology and globalization. All of our efforts to control bees have taught us remarkable lessons, with one consistent outcome being that bees have never been completely controlled. As bees gradually developed resistance to chemicals and mites, genetic research led to new ways of beekeeping.

Society changed too. If the 1950s seemed to be about order and conformity, the 1990s seemed to be the exact opposite. Or were the two decades so different? The arts from this time period used the image of the honey bee to convey a variety of ideas: sex, freedom, power, therapy, salvation, sports and play, love, and finally even home. But through this image, the inarticulate longings of people in transition were quite similar to people in different centuries trying to find form and comfort.

Epilogue

> I am fascinated by the interactions between bees. I am fascinated by the interactions between beekeepers.
>
> —*Marla Spivak*

In Hawaii, Michael Kliks, owner of the Manoa Honey Company, and his assistant Keoki Espiritu have invited me to see their hives on Oahu. We don bee veils and jackets, the garments setting each of us in relief against the smoky cloud-covered mountain. The pink jasmine vines threaten to overtake us. Tall banana trees shade the hives, a colorful combination of mainland and Molokai boxes. Yellow ginger punctuates the green setting like confetti. Together, we clear out the brush to create more space for Kliks's hives. The *false pakaki* (fake jasmine) vines need to be cut back, so we work for a couple of hours. When we finish, there's an orderly little bee yard in the midst of this Hawaiian jungle. We watch the body surfers at Wanamaeigo, munching on Keoki's bologna sandwiches, credit going to his wife.

California. John Miller, a fourth-generation beekeeper descending from N. E. Miller. Ever the native migrant, he's been making the trek from Blackfoot, Idaho, to California since he was fourteen. Add Gackle, North Dakota, to his tour, and, well, there is one of America's last real cowboys—to use Douglas Whynott's phrase—alive and well in the twenty-first century.[1]

Driving down the Sacramento Valley, we see the industrial countryside has indeed thrived: there are almond orchards on one side, being planted in new diamond-shaped patterns as opposed to the 1950s grid square because the new research suggests that diamond patterns allow for two more rows of trees, thus increasing the almond yields. Thus, the traditional pollination formula has been revised in California, so that a grower needs two or three hives for approximately 130 trees. Furthermore, trees are now uprooted at the end of a thirty-year cycle. New planting patterns make life easier for the beekeeper, who must pull the hives in and out of fields. There is more room between the rows to back in and out, a detail most orchard growers don't think about.

Yet once we get into the fields, the Miller crew is as about as individualized as the orchards are systematized. There's nineteen-year-old Mike Asher, who's filling feeders with sugar water from a large tank as casually as if he were watering flowers. There's Layne, a former CEO transitioning into another career. There's Jay, with six daughters, who welcomes me into the crew by reminding me that he is the better looking, smarter, and nicer brother of the family. John—well, he's busy checking the numbers.

Each bee colony is refreshingly different too. One colony needs more sugar, one has mites—no, make that two colonies. Mike calls from a couple of hives over, "This hive has varroa too." I have never seen varroa mites, but the men take time to show me the red specks, almost like drops of blood, on the bees. And I see the damage too: the deformed wings, the slower pace of the bees, the undeveloped workers. The men confer about the next steps: Perhaps we need to do some tests. Let's try some sugar. Okay, let's mark these and finish feeding. These steps are part of a normal day. And when Eric Mussen estimated the value of those days in 2002, he concluded that John, Jay, and Layne Miller and other beekeepers in California "earned $52 million in honey and beeswax, nearly $62 million renting colonies for crop pollination, and $5.5 million selling queens and bulk bees to other beekeepers. Furthermore, the value of the crops grown in California that were pollinated by honey bees was $4,353,460,249." An amount that he says, "eclipses the $126,651,220 total gross earnings of all beekeepers.[2]

Apitourism—tourists that visit other bee yards—has a good future, especially as more novices join the ranks and the beekeeping conferences continue to draw crowds. In surroundings such as Hawaii and California, apitourism can appeal to an emerging niche market in ecotourism. Apitourism also defines a new direction for beekeepers in other equally exotic places in the varied American landscape: the Rio Grande, the Appalachians, the Plains, the East Coast. American beekeepers have always been travelers, and this aspect of the industry acknowledges the insatiable curiosity beekeepers have when visiting other colonies.

Bees in an Electronic Countryside

If I may tinker with Steven Stoll's thesis, America has become the electronic countryside, and beekeepers are very much at the heart of this shift. Information about bees is now digital—DNA nucleotide sequence, research, information, e-mail, and Cornell University's digital bee library.

Apitourist that I am, I have great difficulty remembering that honey bees are being used to detect bombs and chemical weapon. Yet this research represents the latest development in honey bees. The Defense Department's Advanced Research Projects Agency (DARPA) had been conducting tests in which honey bees were used to track land mines in Afghanistan. Because honey bees can pick up trace amounts of chemicals in the flying zones, they are good indicators of major changes in the terrain.

Honey bees also can be trained to fly to certain scents. According to journalist Andrew Revkin, "Scientists working for the Pentagon have trained ordinary honey bees to ignore flowers and home in on minute traces of explosives, a preliminary step toward creating a detection system that could be used to find truck bombs, land mines and other hidden explosives."[3] Alan Rudolph states, "even relatively simple organisms can boast sophisticated abilities to sense their environments and move about in them."[4]

This project is in its early stages, but according to Rudolph, "over the past eight months, researchers have mimicked land mines' 'smell' in the field. The bees' ability to locate targets is impressive." Bee researcher Jerry Bromenshenk provides further analysis of bees' olfactory senses. The Bee Alert team, stationed at the University of Montana, has concluded that bees can be trained to use their olfactory senses to detect chemicals with little or no reward such as sugar or honey. In other words, bees can be trained to explore for certain scents if done very early, before they link sight with taste.

There are downsides to the project. Honey bees have time constraints: "Bees, like dogs, have limitations," writes Revkin; "they do not work at night or in inclement weather." Furthermore, when the World Trade Center was bombed on September 11, 2001, the entire project was suspended because the Defense Department underwent radical infrastructure changes.

There is, within the commercial beekeeping community, this phrase: "the unknown transfer of ownership of hives." *Theft* has such negative overtones. However, it does happen, although especially in the California desert, hiding such a "transfer" is difficult, and generally, a commercial beekeeper can "transfer" the hives back to the business. In other areas, many beekeepers can lose their hives and never recoup their losses.

So Jerry Bromenshenk and Bee Alert are exploring the possibility of electronic hives. The concept is very similar to the branding of hives, except that electronic tools are used. If an electronic bar-coded tag is registered with the local authorities and then placed into the hive, local authorities can trace hives by doing a simple scan, as happens at many grocery stores. Migratory beekeepers take care of their hives by using computers to see which hives need immediate assistance or which ones have been stolen or vandalized. With this technology, beekeepers can monitor many hives from their homes. In this regard, migratory beekeeping may become—dare we say it?—less migratory.

Speaking of electronics, the Internet has provided many opportunities to educate, connect, market, and communicate. Some bee journals are now available online. Information about conferences is at fingertips. Extension specialists are available via e-mail as well as by phone. And communities that previously shunned contact with the secular world, such as Trappist monks, have an avenue to market their products while staying true to their monastic codes. Libraries have gone digital, providing many of this industry's foundation texts online.

In addition, researchers Dr. Sunil Nakrani and Craig Tovey are just beginning research that compares the similarities between honey bee dances and electronic networks. At first glance, one may miss the link between two such different fields. But, just as forager bees communicate information about floral sources to other bees, so too do Internet servers communicate information to customers.

Put simply, if all forager bees spent too much time tracing poor sources of nectar, the entire colony would suffer. Hence, the bees have adapted the waggle dance, which clearly communicates the quality and distance of the food source. The longer the dance in the bee world, the poorer the food source in terms of quality: "A nectar-collecting bee judges how good its flower patch is . . . by seeing how long it takes to find an unemployed food-storer bee. If it takes ages, then the forager concludes that the patch is nothing special. . . . But, if there is a plethora of food-storer bees, then the forager realizes that it has struck lucky."[5]

Similarly, the longer a customer waits for information on the Internet, the more one pays in time and money. So, in order to increase the efficiency of computer servers, Nakrani and Tovey have adapted algorithms to measure the similar difficulties between the bee dances and electronic servers to see if there are ways that computer programs can eliminate long searches. It seems that Americans continue to learn how to be more efficient from honey bees, even when it comes to computers!

The Honey Bee Genome Project may affect how beekeepers can work with African bees and treat their bees for destructive diseases such as foulbrood and mites. The project began with a desire to help advance multiple fields, such as molecular biology, biomedicine, and agriculture. Knowledge of the genome may help breed bees that are more resistant to varroa, may help researchers develop gentler strains of African bees, and may help those allergic to bee stings. Because the infrastructure for studying the human genome project was already in place, Dr. Gene Robinson was able to pursue his goals rapidly. The U.S. Department of Agriculture teamed up with the National Human Genome Research Institute to donate $7 million to the project. Although the project was only approved in 2002, it was completed at Baylor University in 2003. Having such a sequence, Robinson explained in an interview, is a "ticket to twenty first century technology. It does not provide a cure to the problems of beekeeping, but it does provide a foundation to address these issues from a new way."[6]

Ecological studies are teaching us more about the relationship between bees and plants in the Southwest. A big discovery in 2001 was that mistletoe, long respected for its powers to encourage romance, may be a food for bees in the Southwest. "The plant is a parasite that attaches itself to trees. But when researchers recently looked at areas where mistletoe had run amok," reports Seth Borenstein, "they found increased populations of bees and birds that feed on the plant's whitish berries."[7] In fact, mistletoe blooms in the Southwest in February, a season in which there are very few pollen-producing plants in bloom. "Mistletoe has more of a role than we might imagine in our ecosystem," said USDA bee researcher Eric Erickson, in Tucson, Arizona.[8]

Migratory Patterns in the New Millennium

Although it is too soon to tell how the African honey bee (AHB) will affect the major pollination centers such as California and Florida, researchers have been surprised at how slowly and contained the African bees have been since they arrived in Hidalgo, Texas. Experts had predicted that African honey bees would have spread across the southern region of the United States. However, as of 2004, the bees are in southern California, Arizona, New Mexico, Nevada, Oklahoma, Kansas, and Texas as well as Puerto Rico and the Virgin Islands. The other southern states—Louisiana, Alabama, Mississippi, and Florida—seem to have a natural barrier caused by humidity, states USDA entomologist Jose D. Villa in an article written by Kim Kaplan: "What immediately jumped out at me was the correlation in rainfall," Villa explains. "Rainfall over 55 inches, distributed evenly throughout the year, is almost a complete barrier to the AHB spread."[9] However, Villa does not know exactly why rainfall might be limiting AHB dispersal. Perhaps the best approach in dealing with the mystery of the African honey bees is Spivak's lesson when she was working in Peru and Costa Rica. "The Latin American counties know how to take it as it comes."[10]

As the Latin American countries have discovered, beekeeping will not be the same. Marla Spivak explains that beekeepers will need to leave more space between hives and approach hives without provoking the insects; researchers need to consider more genetic breeding of the gentler strains; and civilians and hobbyists will need to invest in protective gear. *Bee Culture* editor Kim Flottum estimates that more hobby beekeepers will leave the bees to commercial beekeepers. On the other hand, a side benefit of African bees is that people are organizing conventions, learning about the bees, and preparing for the species well in advance of the swarms.

Quarantines mandate that beekeepers may enter a quarantined zone, but once in, they must keep their bees in that place—or try to. Bees often swarm, though, so the quarantines can prove ineffective, although they slow advances of threats such as mites or African bees. The quarantine in Texas, designed to slow the African honey bees' advance, is on a county-by-county basis, which beekeepers respect, but African bees don't. However, beekeepers confined to a quarantined zone rich in nectar and pollen stand to make handsome gains in terms of pollination fees, if their bees are healthy.

Or losses, if one looks at the counterargument. One lesson about the North Dakota quarantine of 1986 is clear: beekeepers have gone bankrupt because of decisions made by uninformed politicians. North Dakota winters aren't conducive to large commercial operations left stranded. There is worry that the quarantines will reduce the hobbyists. If hobbyists decline in number, the pool of experienced people who can handle a bee-related emergency is substantially reduced. If a truck were to overturn in Medina, Ohio, Kim Flottum has a list of hobbyists he can organize to respond to such an emergency. But if the quarantines stay in effect, Flottum predicts, the number of people with equipment and expertise to handle bees will drop substantially.[11]

Beekeepers continue the centuries-old pattern of immigration and emigration, bringing knowledge about other species and skills from other continents. A case in point: Robin Mountain, a second-generation beekeeper, first began working with African bees in Pretoria, South Africa. When he immigrated to California, he worked in the commercial queen and package bee industry in California and Texas. He brings to the Ohio Valley a combination of extension work and genetic research. While organizing a portable extracting unit in conjunction with Kelley's Bee Supply, Mountain and his family are training a new group of queen producers in the Kentucky region. The dual nature of his work has one purpose: "We're trying to create a farming industry," he explained to Jason Simpson of the *Leitchfield Record*.[12] Kentucky is a far cry from the industrial countryside that is California, but for now, beekeeping cooperatives can rent the portable extracting unit for affordable prices. For the first time, Appalachian beekeepers have access to equipment they wouldn't ordinarily have.

Mountain also encourages beekeepers to rear their own queens selected for the unique regions in the Ohio Valley. In "Rearing Your Own Kentucky Queen," Mountain explains that "probably the most important reason beekeepers leave the specialization of queens to a specialist is that the specialist can produce them at less expense than could other beekeepers."[13] They have the specialized equipment, the labor force, and the established shipping lines. As part of his extension work, Mountain teaches beekeepers that they no longer have to accept this set of circumstances. If beekeepers want to develop a bee adapted to winters or with more disease resistance, or even with a gentler temperament, they can. This possibility is exciting for beekeepers who have always depended on the queen and package bee businesses located in the West or the South. It remains to be seen how Mountain's commercial expertise will affect this region, but it is one small example of how important extension work has become in this century.

The one group that stands in stark contrast to the migratory beekeepers is the Amish, who have been beekeepers since they first set foot on this continent. The four inventions from the nineteenth century still serve their needs just fine. Rather than upgrade to powered extractors, still use hand-cranked versions. "They keep their hives clean," explains former bee inspector Stephen Henderson. "They don't engage in migratory travel, so they don't run the same risks of bee diseases."[14]

That's just it. When the commercial beekeepers chose to follow the bloom in the 1900s, they were following an American value system that rewards efficiency and ingenuity. The industrial countryside had not counted on the varroa and tracheal mites in the 1980s, but it does seem as if the beekeeping industry is working with scientists to develop ways that bees can protect themselves. Marla Spivak speaks with great conviction regarding her research: "My purpose was not to create a new line [of honey bees]. It saddens me that a hive cannot protect itself. My primary goal is to wean bees and beekeepers off chemicals."[15]

The Amish choose not to follow the bloom. In so doing, they sidestep many issues facing beekeepers: no federal quarantines, state and federal inspections, machines. To quote Steven Stoll, "The Amish hold these values above all others: anything that undermines their ability to cohere as a community of neighbors and linked families, anything that isolates them in their work or places production for profit ahead of the collective process, is prohibited."[16]

Although some use chemicals and gas-powered air compressors to operate extractors, the Amish tend to be affected by common problems a little differently than others, perhaps a little later, perhaps using a different type of cure. For instance, mites and diseases spread rapidly because of interstate travel, but the Amish tended to be affected later than everyone else. But this doesn't make them throwbacks to the nineteenth century. Conversely, they have a vested interest in the new studies regarding hygienic behavior, suppressing mite resistance, and developing queens suited for their particular regions. Stoll argues that the Amish, by maintaining a sustainable system of agriculture from the 1820s, "represents an alternative—a progressive occupancy of land for the twenty-first century."

In a land not unified by language, religion, or political party, the honey bee still serves as a metaphor for stability and social order in America. The Utah State Capitol has kept the 1980 Grand Beehive exhibit on permanent display. Newspapers have incorporated the hive into featured drawings.[17] Television commercials have continued to use bees to market everything from insurance to honey. Even though beekeeper Allen Cosnow wrote in a letter to the *American Bee Journal* in 2003 that he'd like to see the skep eliminated, the chances of that happening in the near future do not seem likely.[18]

Although the number of minority beekeepers remains very small in the nation, Sue Monk Kidd's *The Secret Life of Bees* explores in a fictional narrative three black women who were beekeepers during the 1960s. In the novel, three black beekeepers—May, June, and August—take in a young girl named Lily Owens and her black caretaker, Rosaleen. The black beekeepers market their honey using a black Madonna. It is this image to which Lily, abandoned by her mother, is instinctively drawn. When asked about her unusual label, August replies, "Everybody needs a God who looks like them."[19]

An adolescent, Lily is trying to figure out her troubled past, but more to the point, she's trying to figure out how to live in the future. By the end of the novel, Lily sees that a "cluster of beehives is the eighth wonder of the world." "From a distance," Kidd writes, "it will look like a big painting you might see in a museum, but museums can't capture the sound. Fifty feet away, you will hear it, a humming that sounds like it came from another planet. . . . Your head will say don't go any further, but your heart will send you straight into the hum, where you will be swallowed by it. You will stand there and think, I am in the center of the universe, where everything is sung to life."[20]

Although the novel is filtered through Lily's eyes, Kidd positions black women in the beekeeping community, which has not been done since Frank Pellett was writing about the South in the 1920s. Kidd extends a pastoral tradition begun with Zora Neale Hurston's *Their Eyes Were Watching God* in the 1930s. Kidd reminds her readers of the fundamental promise that the American Dream is open to everyone, including black women, and the land will provide if one is willing to learn its complexities.

Through the course of writing this book, I have come to agree with G. H. Cale, writing in 1943, that honey bees and their keepers have become stepchildren in United States agriculture. The twentieth century, with its emphasis on industry and technology, created a support system between farmers, extension specialists, and the federal government to respond to agricultural emergencies. But, beekeepers still seem to be on the margins of this infrastructure. We need to strengthen those systems as our country enters a new millennium. We cannot continue to take for granted honey bees or their keepers.

We also need to rethink our long-held associations between bees and people—specifically poor people. The link between drones and the poor has been weakening since the 1900s, but if America is to continue to be a model hive, we have to take better care of our workers in the new millennium by offering better insurance, better retirement, flexible schedules, and job security. As a society, America can do better when it comes to offering better labor conditions, better communication patterns, and more equitable power distributions for its workers. How do I know this? Nineteenth-century business owners Charles Dadant and A. I. Root established businesses in which workers were given progressive benefits; both businesses still influence the industry in 2004. In the twentieth century, Walter Kelley bequeathed his company to his employees. We can take our cue from beekeepers, who know quite well the importance of proper management of workers.

✾ *Epilogue* ✾

My last argument in this book has been that globalization was not just a twentieth-century political term, but rather a phenomenon that has happened among beekeepers in this country for four centuries. This book has been a study in analyzing how this little insect pulls together diverse societies within our borders. In their love for the bee, American beekeepers and honey hunters have crossed—and I predict will continue to cross—racial, gender, ethnic, religious, and education differences that can separate other groups. As import trade continues to become a factor in domestic honey trade and the 1992 Honey Bee Restriction Act has been relaxed, I look forward to remaining fascinated by the interactions between bees and by the interactions among beekeepers.

Notes

Introduction

1. Charles Hogue, "Cultural Entomology," 192. Hogue advocates more research be done to further an emerging and valid field of knowledge he calls "cultural entomology." The first colloquium on cultural entomology took place at the Seventeenth International Congress of Entomology in Hamburg in 1984.

2. Gene Kritsky, "Castle Beekeeping," 26.

3. Charles Butler, *The Feminine Monarchie: Or a Treatise Concerning Bees and the Due Ordering of Them* (1609). For a discussion about gender politics that existed *before* Butler's book was published, see Jeffrey Merrick, "Royal Bees." He argues that "the hive was not a symbol of female authority to many of the French regents such as Catherine de Medici, Marie de Medici, Anne of Austria or Elizabeth I" (15).

4. Kevin Sharpe, *Politics and Ideas in Early Stuart England,* 54. Sharpe provides further information about the spiritual nature of Butler's writings, which instructed all members of the commonwealth to pay tithes. Sharpe argues that subsequent writings about honey bees became more secular after the 1630s.

5. Frederick Prete, "Can Females Rule the Hive?" 125. Prete's wonderful article traces British beekeeping texts and their audiences until the eighteenth century.

6. Carl Bridenbaugh, *Vexed and Troubled Englishmen,* 95.

7. Mark Kishlansky, *Monarchy Transformed,* 29.

8. Peter Linebaugh and Marcus Rediker, *The Many-Headed Hydra,* 39.

9. Karen Ordahl Kupperman, "Beehive as a Model." I'm grateful to Kupperman for this thoroughly researched article on bees and the social model that the English transferred to the colonies.

10. Tom Webster, interview with the author, April 2, 2004.

11. Kupperman, "Beehive as a Model," 273.

12. Ibid., 286.

13. Timothy Raylor, "Samuel Hartlib and the Commonwealth of Bees," 96.

14. Ibid., 117. Incidentally, Hartlib's best friend was a victim of the famine in 1840s. He literally starved to death.

15. Kupperman, "Beehive as a Model," 288.

16. Karen Ordahl Kupperman, "Death and Apathy."

17. Kupperman, "Beehive as a Model," 273.

18. E. W. Teale, "Knothole Cavern."

Chapter 1. Bees and New World Colonialism

1. George Peckham, "A True Report of the Late Discoveries," in *Envisioning America,* 64.

2. The English were not the first to bring cattle to North America. The Spanish had imported cattle into the area that is now Texas in the 1500s. No record exists that they tried to bring honey bees, although they did profit from the stingless bee that produces honey in Mexico. This variety is different from the *Apis mellifera,* however.

3. Eva Crane, *World History of Beekeeping and Honey Hunting.* The *Sea Venture* was the flagship of the Virginia Company of London, and when the English decided to take possession of the Bermudas, other colonists arrived, including Robert Rich. Crane provides documentation of a letter that Rich wrote to his brother, Nathaniel, dated May 25, 1617, from the Bermudas: "The bees that you send doe prosper very well" (358). Crane also provides a footnote that the wreck of the *Sea Venture* provided the theme for William Shakespeare's *The Tempest.*

4. Kupperman, "Death and Apathy," 24.

5. Kupperman, "Beehive as a Model," 278.

6. Ibid., 277. In her article "Death and Apathy in Early Jamestown," Kupperman records that there was the belief that scurvy "attacks only the soft and lazy, and there was, therefore, a moral element in contracting the disease" (34). She doesn't analyze the reasons for this moral judgment, but I think she sets the foundation for an argument that the English were preconditioned to associate inactivity with dronelike behavior.

7. Crane, *World History of Beekeeping and Honey Hunting,* 358–59. Crane lists the captains and names of four ships from the Virginia Company that had transported the items on November 21, 1621: John Huddleston of the *Bona Nova,* Thomas Smith of the *Hopewell,* Danielle Gale (or Gate) of the *Darling,* and Captain Thomas Jones of the *Discovery.*

8. Kupperman, "Death and Apathy," 36.

9. Kupperman, "Beehive as a Model," 283.

10. William Sachs, "Business of Colonization," 6.

11. Kupperman, "Beehive as a Model," 281.

12. Ibid., 284.

13. William Wood, *New England's Prospect,* 68:54.

14. Sachs, "Business of Colonization," 9.

15. Kupperman, "Beehive as a Model," 282.

16. Laurel Thatcher Ulrich, *Good Wives,* 37; Firth Haring Fabend, *A Dutch Family in the Middle Colonies.*

17. David Freeman Hawke, *Everyday Life in Early America,* 63.

18. Kupperman, "Beehive as a Model," 283.

19. Sachs, "Business of Colonization," 11.

20. Ibid., 13.

21. John Goodwin, *The Pilgrim Republic,* 510–12. Blackstone, who also signed his name Blaxton, was an eccentric hermit who planted a large orchard in the area that eventually became the Boston Common. The type of apple is not known, except that writers referred to it as being a "yellow sweeting." Later in life, after he had married and was coaxed back into preaching, Blackstone would toss golden apples to children in attendance. This trick and the fact that Blackstone would ride to sermons which often were held underneath a tree, on a bull in lieu of a horse would bring spectators to his church.

22. Crane, *World History of Beekeeping and Honey Hunting,* 303.

23. Gene Kritsky, "Advances in Early Beekeeping," 311.

24. Ibid.

25. "Sense and Nonsense: From *ABJ* 1866."

26. An interesting fact is that Lorenzo Langstroth, who revolutionized beekeeping, had ancestors who were part of this movement.

27. Frederick Hahman, "A History of Bee Culture in Pennsylvania," 409.

28. Terry Jordan and Matti Kaups, *American Backwoods Frontier.* Jordan and Kaups write about the emerging Finnish immigrants. As the Finns migrated to Sweden, they gradually encountered problems. Too much woodland and game were destroyed by the newcomers, too much summer pasture and wild meadow were lost by the valley folk, and too much rye from the prolific burned clearings flooded the markets. The resulting conflict, as well as news of another frontier where their shifting cultivation could be freely practiced, brought some Finns to colonial America.

The available evidence, though fragmentary, strongly suggests that the proportion of Finns in the population of New Sweden increased steadily after the mid-1640s. At first the Finnish migration was forced, involving deportations for crimes committed in Sweden, but the early exiles sent back glowing reports to their kinfolk. By the late 1640s, hundreds of Finns had petitioned for the right to emigrate to New Sweden. They began flocking to the Delaware Valley in the early 1650s, and "America fever" raged in the Finnish of Scandinavia. By the end of Swedish rule, at least one-third of the Delaware Valley population was Finnish.

29. John Wuorinen, *The Finns on the Delaware.*

30. Hahman, "A History of Bee Culture in Pennsylvania," 409. Hahman cites a clipping published in the *Altoona Tribune,* owned and edited by an Hon. Henry W. Shoemaker of McElhaton, Clinton County, Pennsylvania.

31. Ibid.

32. Jordan and Kaups, *American Backwoods Frontier*, 59.

33. Ibid., 232.

34. Gerald Francis De Jong, *Dutch in America,* 5. De Jong argues that "it was in the field of commerce that the Dutch distinguished themselves during the seventeenth century. They founded colonies in Brazil, the West Indies, North America, Africa, Ceylon, the Levant, Russia, Japan, and especially the Baltic."

35. Oliver Rink, *Holland on the Hudson,* 171.

36. See P. W. Bidwell, *History of Northern Agriculture;* Crane, *World History of Beekeeping and Honey Hunting;* Bernie Hayes, "The Honey Bee and Napolean Bonaparte."

37. Crane, *World History of Beekeeping and Honey Hunting,* 304.

38. Fabend, *A Dutch Family in the Middle Colonies.*

39. Hawke, *Everyday Life in Early America*, 25.

40. Alfred Brophy, "Quaker Bibliographic World," 245. An aside: Pastorius was one of four men who signed the 1688 "Germantown Protest," which was a formal protest of slavery in Pennsylvania. Even though all four men were highly regarded, their effort did little good. It was another hundred years before Pennsylvania would abolish slavery.

41. Rollin Moseley, "Why the Quilting 'Bee,'" *Bee Culture* 112, no. 12 (1984).

42. Helen Adams Amerman, "Quilting Bee in New Amsterdam," 5.

43. Washington Irving, *Knickerbocker's History of New York,* 353.

44. Although early blacks enjoyed the same rights and privileges as white English settlers, by the 1640s colonial legislatures had enacted laws that gave landowners the right to own chattel slaves.

45. Carl Seyffert, *Biene und Honig im Volksleben der Afrikaner* ("Beekeeping in Africa"), 11. According to linguistics professor Lowell Bouma, the African word "kandir," which is similar to "candle," supports David Livingstone's thesis that the primary market for beeswax in seventeenth-century Africa was the Portugese, who needed it for religious rituals.

46. Crane, *World History of Beekeeping and Honey Hunting,* 529.

47. Seyffert, *Beekeeping in Africa,* 193. In his book, Seyffert considers his own research incomplete, and much of his information was gathered before World War I. As a consequence of WWI, his book was not published until 1930.

48. Ibid., 129.

49. Ibid., 121.

50. Ibid., 130.

51. Neil Salisbury, "The Indians' Old World," 11.

52. Stephen Potter, *Commoners, Tribute, and Chiefs,* 33.

53. Hahman, "A History of Bee Culture in Pennsylvania," 409.

54. Frank C. Pellett, *History of American Beekeeping,* 3.

55. Karen Ordahl Kupperman, *Facing Off,* 161.

56. Alfred Crosby, *Ecological Imperialism,* 188.

57. Everett Oertel, "Bicentennial Bees, Part I," 71.

58. Crane, *World History of Beekeeping and Honey Hunting,* 303.

59. Edward Goodell, "Bees by Sailing Ship and Covered Wagon."

60. In talking about sugar in England, Timothy Raylor discusses Samuel Hartlib's thesis on sugar's possibility to lift England out of a recession. Hartlib concluded that sugar was a problematic commodity for the English to count on in economic terms: it could not be grown in England, it was expensive to produce, and it would do little to alleviate the English economy in 1650 (a time of recession). Thus, Hartlib began his treatise on honey.

61. Everett Oertel, "Bicentennial Bees, Part I," *American Bee Journal* 116, no. 2 (1976): 70.

62. Jordan and Kaups, *American Backwoods Frontier,* 231.

63. Crane, *World History of Beekeeping and Honey Hunting,* 304.

64. Hartlib's *The Reformed Commonwealth of Bees* (London, 1655) contains one of the first architectural designs by Christopher Wren (Timothy Raylor, "Samuel Hartlib and the Commonwealth of Bees," 92). It was one of the first multistoried beehive designs to appear in print.

65. Ann Fairfax Withington, "Republican Bees," 55.

66. Eric Nelson, "History of Beekeeping in the United States," 2.

67. J. Earl Massey, *Early Money Substitutes,* 15.

68. Nelson, "History of Beekeeping in the United States," 2. John Eales was employed by Newbury in 1644 to build beehives. See also Oertel, "Bicentennial Bees, Part II," 138, and S. Bagster, *Management of Bees* (1838).

69. Massey, *Early Money Substitutes,* 16.

70. Arnold and Connie Krochmal, "Origins of United States Beekeeping," 162.

71. Crane, *World History of Beekeeping and Honey Hunting,* 304–5.

72. "To an extent, their advance was due to human intervention, humans with hives on their rafts and wagons moving into Indian territory, but in most cases the avant-garde of these Old World insects moved west independently. . . . But the Appalachians were the real barrier for them. . . . They did get across and then seem to have spread even more rapidly in the Mississippi basin than they had east of the Appalachians" (Alfred Crosby, *Ecological Imperialism*, 189).

73. Eva Crane, *Archaeology of Beekeeping*, 95.

Chapter 2. Bees and the Revolution

1. Lester C. Olson, *Emblems of American Community*, 7.

2. William Eaton, "Sketches from Kentucky's Beekeeping History," 22. Eaton divides Kentucky beekeeping into four periods, beginning with the years 1780 to 1871.

3. Jerard and Pat Jordan, *Spirit of America in Country Furniture*.

4. William Sachs, "Business of Colonization," 13.

5. Everett Oertel, "Bicentennial Bees, Part I," 71.

6. Ibid.

7. Oertel, "Honey Bees Taken to Cuba," 880. This brief article contains a wealth of information. J. Preston Moore was the translator of Antonio de Ulloa's *Noticias Americanas, entretenimiento fisico-historico sobre la America meridional y la septentrional* (translated as *Revolt in Louisiana*, and originally published in Spanish in 1776).

8. Ibid., 880. Ulloa's writings also reflect the common misperception during that time period that bees could injure sugar cane plants. It was assumed that because bees had stingers they could puncture fruit to get nectar. The Spanish were not the only ones to suspect bees of damaging crops. Both German-American and French-American neighbors used to accuse local beekeepers of property damage when grapes were punctured.

9. Lester Breininger, "Beekeeping and Bee Lore in Pennsylvania," 35. An interesting side note is that Breininger cites records from his ancestors for his arguments.

10. Ibid.

11. Wayland Hand, "Anglo-American Folk Belief and Custom," 151, 152.

12. J. E. Crane, "Siftings," 714. Crane uses his own experience in which a beekeeper had died, and thereafter the bee yard had been neglected. "Surely it was a sorry day for that yard of bees when their master died, and they had to pass the winter with so little care."

13. Breininger, "Beekeeping and Bee Lore in Pennsylvania." Breininger uses the phrase "*wann en leicht fatt geet, mus mir die eama ricka.*"

14. Hand, "Anglo-American Folk Belief and Custom," 150.

15. Kathleen Giraldi, "Noah Atwater and His Contribution to Westfield Life."

16. Breininger, "Beekeeping and Bee Lore in Pennsylvania," 39.

17. L. R. Stewart, "Early Midwest Beekeeping," 412.

18. Daniel McKinley, "The White Man's Fly on the Frontier," 444.

19. Hunter James, *Quiet People of the Land,* 13.

20. Ibid., 17.

21. Ibid.

22. John R Weinlick, *Count Zinzendorf.* David Zeisberger was born in Zauchtenthal in the eastern part of Moravia in 1721. When his parents moved to Herrnhut, located on Count Zinzendorf's estate, they were searching for religious freedom. They later moved to Georgia but left their son David in Herrnhut. David then followed his parents to Georgia. While there, David was influenced by Bishop Peter Boehler, the German clergyman who also converted John Wesley, leader of the Methodists.

23. Ibid., 160. An interesting aside is that Count Zinzendorf named Bethlehem, Pennsylvania (which was founded by the Moravians), on his visit to America in 1741. According to Weinlick, "Only a Zinzendorf could have christened a community so dramatically. With the Brethren gathered in a barn housing themselves and their cattle, Zinzendorf began to lead them in a German Epiphany carol combining Christmas and missionary thoughts."

24. R. Douglas Hurt, *Ohio Frontier,* 87. I'm grateful to Hurt for his book on the Ohio frontier and his sensitive treatment of Zeisberger's involvement with both the British and American forces.

25. Ibid., 91.

26. David Zeisberger, *Diary of David Zeisberger,* 294.

27. Crane, *World History of Beekeeping and Honey Hunting,* 306.

28. Ibid. Crane mentions the Delaware Indians but doesn't frame the context very well. For information about Chief Echpalawchund, see Edmund De Schweinitz, *The Life and Times of David Zeisberger,* 384, 393. De Schweinitz notes that Echpalawchund was a powerful Delaware chief. When he attended the Schonbrunn programs during Christmas 1773, he decided to convert to Christianity. He was christened Peter, and his conversion had powerful implications among the Delaware Indians. Many other Delawares followed his example in January 1774, and Chief Peter helped negotiate a shaky truce between the Delawares and the Moravians, whose position was already precarious because they were located between American Revolutionaries and British Tories.

29. McKinley, "The White Man's Fly on the Frontier," 445.

30. James Mooney, "Myths of the Cherokee," 82.

31. Daniel H. Unser Jr., "Frontier Exchange Economy," 234–35.

32. Crane, *World History of Beekeeping and Honey Hunting,* 261.

33. Ibid.

34. Crane speculates that beekeeping practices migrated to Africa from Egypt, which had once ruled Africa in ancient times: "In Ancient times, Nubia (Cush) adjoined Egypt to the south and extended along the Nile valley from Aswan to Khartoum (now in Sudan) and over neighbouring deserts. . . . So knowledge of hive beekeeping could well have reached what is now Sudan from Ancient Egypt" (*World History of Beekeeping and Honey Hunting,* 261).

35. Rosalie Fellows Bailey, *Pre-Revolutionary Dutch Houses and Families,* 20. "These substantial houses, especially the stone houses, are the result of the abundant and inexpensive slave labor of the times, the Proprietors of New Jersey granting 75 acres for every slave brought into the colony. It is exceedingly doubtful if the stone would have been so extensively quarried, cut, dressed, carried and laid if there had been no slaves. . . . The term 'Dutch colonial house' has often been used for the type of one and a half story house developed here by the Dutch, which flourished to perfection especially in Bergen County, New Jersey. The name is really a misnomer for it came into existence after the fall of the New Netherland government and reached its greatest height in the half century after the American Revolution."

36. *Africans in America.* PBS series. Dir. Orlando Bagwell. 1998.

37. Vincent Carretta, "Olaudah Equiano or Gustavus Vassa?" Carretta provides strong evidence to refute Equiano's narrative in which he claims he was born in Eboe, Guinea (present-day Nigeria), in 1745. Carretta argues that Equiano was actually born in South Carolina and used oral narratives from other slaves to form a picture of preexploration Africa in his slave narrative.

38. Vincent Carretta, *Unchained Voices,* 191.

39. Ann Harman, "Home Harmony," 350. Harman's recipe is taken from *The Honey Book* by Lucille Recht Penner (no further bibliographic information is provided).

40. Antoine Le Page du Pratz, director of a large plantation at Chapitoulas owned by the Company of the Indies, recommended that owners "give a small piece of waste ground" to their slaves, "engage them to cultivate it for their own profit," and purchase their produce upon "fair and just terms" (Daniel H. Unser Jr., "Frontier Exchange Economy," 229).

41. Ibid.

42. King Carlos of Spain did the most in terms of maintaining a steady Spanish presence in Texas and Louisiana. He wanted Louisiana to serve as a buffer zone between the silver mines in New Mexico and the Americans to the east.

43. Unser, "Frontier Exchange Economy," 230.

44. Handbook of Texas Online, "Old San Antonio Road." University of Texas. Available at: http://www.tsha.utexas.edu/handbook/online/articles/view/00/ex04.html.

45. Unser, "Frontier Exchange Economy," 232. Although Louisiana was still technically part of the French empire, Antoine Le Page du Pratz wrote in 1763 that honey bees in Louisiana, which included the lower Mississippi Valley, lodged in nests to secure their honey from bears. If the Choctaw tribes used honey or beeswax in their trades in south Alabama, Mississippi, and Red River regions, however, it would have been after 1771–1772. Unser reports that in 1771–1772, Choctaw camps were occupied with the deerskin trade; they "rarely neglected to exchange venison, bear meat, and tallow for ammunition, cloth, and drink with settlers and travelers whom they encountered during hunting season" ("Frontier Exchange Economy," 231). His research suggests that the Choctaw were unfamiliar with beeswax and its benefits, but conclusive evidence about later dates is lacking.

46. Jean Crèvecoeur, *Letters from an American Farmer,* 32–33.

47. Ibid., 34.

48. Ibid., 32, 131.

49. Ann Withington, "Republican Bees," 44.

50. Robert Garson, "Counting Money," 22.

51. Kenneth Scott, "A British Counterfeiting Press in New York Harbor, 1776," 117. Many thanks to Eric Newman, who called my attention to this article.

52. "How the Bees Saved America," 308.

53. Withington, "Republican Bees." I am indebted to Withington's careful research and the illustrations she provides with her article.

54. Ibid.

55. Bodog F. Beck, "The Great Seal of the State of Utah," 517.

56. Ibid.

57. Arnold and Connie Krochmal, "Origins of United States Beekeeping."

58. Philip A. Mason, "American Bee Books," 37.

59. Ibid.

60. Ibid., 34.

61. Roger and Kathy Hultgren, "Old Swarming Devices," 158.

62. Mason, "American Bee Books," 34.

63. "Bee, n1" *Oxford English Dictionary.* Ed. J. A. Simpson and E. S. C. Weiner. 2nd ed. Oxford: Clarendon Press, 1989. *OED Online.* Oxford University Press. Accessed on December 4, 2003, at http://dictionary.oed.com/cgi/entry/00019338.

Chapter 3. Before Bee Space

1. A. I. Root, *ABC and XYZ.*

2. Kim Flottum, interview with the author, 2004.

3. Wyatt Mangum, interview with the author, December 18, 2003.

4. Frank C. Pellett, *History of American Beekeeping,* 31.

5. Gene Kritsky, "Beekeeping in Glass Jars"; Bagster, *Management of Bees.*

6. Gene Kritsky, "Octagonal Hives," 881.

7. Florence Naile, *America's Master of Bee Culture,* 66.

8. McKinley, "The White Man's Fly on the Frontier," 449. Beeswax was especially important for household uses and frontier trade. In Missouri, beeswax cakes called "yellow boys" were used as currency.

9. These states included Missouri, Wisconsin, Iowa, and Kansas.

10. Frank C. Pellet, *History of American Beekeeping,* 31. But for a more extensive discussion of Kirtland's importance, also see R. M. Rogers, "Jared Potter Kirtland, Amateur of Horticulture," *Journal of Garden History* 6, no. 4 (1986): 357–75.

11. Crane, *World History of Beekeeping and Honey Hunting,* 394.

12. McKinley, "The White Man's Fly on the Frontier," 446.

13. Ibid.

14. Roger Welsch, "Funny Beesness," 36.

15. Washington Irving, *A Tour on the Prairies,* 51.

16. James Keefe and Lynn Morrow, eds., *White River Chronicles.*

17. Ibid., 198.

18. Irving, *A Tour on the Prairies,* 47.

19. Ibid., 50.

20. Ibid., 52.

21. J. Frank Dobie, "Honey in the Rock," 125.

22. Bodog F. Beck, "Tom Owen, the Mighty Honey Hunter," 162.

23. Walter Webb, *The Great Plains* (1931), 94–126. The Spanish conquistadors underestimated the Plains Indians, who were a force to be reckoned with. In terms of conquering a nation, the Spanish had a four-step system: the conquistador, or military representative, was the advance system, followed closely by those who would establish pueblos, presidios, and missions. In other words, the rudiments of civilization worked hand-in-hand with the conquistador—and until the Spanish arrived in Texas, this system worked quite well. However, in terms of matching the conquistadors with the Native Americans, the Spanish were no match for the Apache, Comanche, or the cannabilistic Tonkawas. The irony was that the Spanish introduced horses to the Plains, which became the Indians' best weapon against the Spanish. Furthermore, the distance between

missions and pueblos was so great that the Spanish gave up trying to convert the Indians.

24. McKinley, "The White Man's Fly on the Frontier," 447.

25. Ibid.

26. Ibid., 448.

27. Ibid., 450.

28. Irving, *A Tour on the Prairies,* 54.

29. Barrels for liquids such as honey were generally made of hard woods like oak. The most secure barrels would have six rings rather than four because the barrels needed the tension provided by the rings, according to Middleton Plantation volunteer, Doug Nesbitt, a volunteer cooper who demonstrated the art of barrel making on May 24, 2003. For more information about barrels, see Ken Olson, "Barreling Honey 100 Years Ago," 518.

30. McKinley, "The White Man's Fly on the Frontier," 450.

31. Dobie, "Honey in the Rock," 124.

32. The section heading is taken from the old Greek saying, "It is sweet and good to die for one's country." This episode is taken from Sue Hubbell, "Honey War," 181.

33. Hubbell, "Honey War," 187–88.

34. McKinley, "The White Man's Fly on the Frontier," 446.

35. Christian Goodwillie, *Shaker Songs,* 24.

36. Ibid., 16.

37. Julia Neal, *Kentucky Shakers,* 19; Goodwillie, *Shaker Songs.*

38. Isaac Newton Youngs, *Journal: Tour with Brother Rufus Bishop.* I am grateful to Christian Goodwillie, curator at Hancock Shaker Village, who supplied this information.

39. Giles B. Avery, "A Journal concerning Bees in the Second Order." Unpublished journal, Hancock Shaker Village, Pittsfield, Mass., 1851–54. Credit goes to Brian Thompson and Larrie Curry at Pleasant Hill Shaker Village for providing a copy of this journal to me. In addition, both provided more general information about the Shakers.

40. Daniel Patterson, *The Shaker Spiritual,* 223.

41. Donald Pitzer and Josephine Elliot, "New Harmony's First Utopians."

42. Donald Carmony and Josephine Elliot, "New Harmony, Indiana."

43. Joseph Moffett, "Mitchell Brothers of Missoula, Montana," 508.

44. Bodog F. Beck, "The Great Seal of the State of Utah," 517.

45. M. G. Dadant, *Life of C. P. Dadant,* 12.

46. Viktor Bracht, *Texas in 1848,* 104.

47. Webb, *The Great Plains,* 184–202. Walter Prescott Webb explains the "hidden" agenda behind the Gadsden Purchase: Gadsden and Davis really wanted

to extend the slave-holding empire of the South. It was thought that if a railroad could connect the South from the Carolinas to California, then slavery could be a firmly entrenched institution, regardless of Congress.

48. Bracht, *Texas in 1848,* 109.

49. Rudolf Leopold Beisele, *History of the German Settlements in Texas.* The Adelsverein, the company that brought people to Texas, went bankrupt in 1847 because of the poor land choices made for its people. Among other things, it bought land that belonged to the Comanche Indians. According to the Sophienburg exhibit of daily life at the New Braunfels Archives and Museum of History, the Adelsverein was a "high minded but poorly informed society that engaged in questionable business deals."

50. Bracht, *Texas in 1848,* 96.

51. Clark Griffith Dumas, "Apiculture in Early Texas."

52. Joseph Moffett, *Some Beekeepers and Associates.* Diehnalt's descendants have stayed in the business and even run a museum!

53. Cecil Woodham-Smith, *The Great Hunger.*

54. Camp Ezell, *Historical Story of Bee County, Texas.*

55. However, Bee County was named after Sam Houston's second in command.

56. Meridel Le Sueur, *North Star Country,* 129.

57. Cobbett was writing to Morris Birkbeck, Esq. The letter can be found in *A Year's Residence in the United States of America,* 291. I'm grateful to Steven Stoll for bringing this letter to my attention.

58. Stephanie Coontz, *The Way We Never Were.* See in particular the chapter entitled, "We Stood On Our Own Two Feet."

59. Records exist that suggest that as many as thirty-eight manuals were printed in the United States by this time, but nothing in Langstroth's writings indicate that he was aware of them.

60. Gene Kritsky, "Langstroth and the Origin of the 'Bee Space,'" 813. F. R. Cheshire, *Bees and Beekeeping* (1888). James Heddon introduced the term "bee space" in 1885, although Langstroth had patented the idea in 1852; Langstroth was very careful not to reveal details in case of copyright infringement.

61. Naile, *America's Master of Bee Culture,* 43.

62. Ibid. See also Frank C. Pellett, *History of American Beekeeping,* 9. The sources are still unclear on what types of headaches Langstroth suffered, but they were so debilitating that he could not carry out his pastoral duties, although he hated to give up the ministry and tried to return on several occasions.

63. Naile, *America's Master of Bee Culture.* This respect for minorities and women also seemed to be passed down from his ancestors. His great-grandmother Elizabeth Lorraine Dunn taught her slaves to read when it was illegal to do so and freed them once she became convinced of the evils of sla-

very. "The once wealthy widow thus deprived herself of all but a modest competence" (35).

64. Ibid., 64–65.

65. Crane, *World History of Beekeeping and Honey Hunting,* 382.

66. Kritsky, "Langstroth and the Origin of the 'Bee Space,'" 811. A. J. Cook, *The Bee-Keeper's Guide or Manual of the Apiary.*

67. Kritsky, "Langstroth and the Origin of the 'Bee Space.'"

68. Naile, *America's Master of Bee Culture,* 113.

69. Crane, *World History of Beekeeping and Honey Hunting,* 371.

70. Sue Hubbell, *Book of Bees,* 71.

71. Mason, "American Bee Books," 484.

72. Frank C. Pellett, *History of American Beekeeping,* 163.

73. Hal Cannon, telephone interview with the author, 2002.

74. Mark Twain, *Roughing It,* 90.

75. Carol Simon, "Utah's Busy Beehives." The Mormons created an entirely new language based on phonetics in order to help immigrant converts assimilate easily into American culture. But the English-speaking faction won out, and thus, even though the Mormon alphabet is still printed, it is not spoken.

76. Thomas Cheney, *Mormon Songs from the Rocky Mountains,* 109.

77. Hal Cannon, *Grand Beehive.*

78. Wallace Stegner, *Mormon Country,* 20.

79. Moffett, "Mitchell Brothers of Missoula, Montana," 508.

80. Cannon, *Grand Beehive.* Having already served with distinction in the Mexican War, Joseph Johnston later served in the Civil War as a general for the Confederate forces.

81. Thomas L. Kane, "The Mormons" (speech, Pennsylvania Historical Society, Pennsylvania, March 26, 1850).

82. Webb, *The Great Plains.*

83. Lee Watkins, "John S. Harbison," 241–42.

84. Ibid., 244. According to Watkins, "It is ironic that he [Harbison] designed this hive so that the bees would be better able to utilize their stores during severe winters since there were no such winters in California, especially not San Diego County, where Harbison eventually migrated and produced his world record honey crops."

85. Stephen Van Wormer, "Beeves and Bees," 50.

86. Catherine Williams, "Bringing Honey to the Land of Milk and Honey," 33. This is the same time period that Langstroth published his work, Longfellow published *Hiawatha,* and Harbison successfully transported his bees to California.

87. Ibid.

88. Ibid., 36.

89. Bruchac, "Is *Hiawatha* an Authentic Native American Story?" afterword in *Hiawatha and Megissogwon*, n.p.

90. Henry Wadsworth Longfellow, "The Song of Hiawatha," *Hiawatha*, book 21, 272.

91. Joseph Bruchac, "Is *Hiawatha* an Authentic Native American Story?" In *Hiawatha and Megissogwon*. In his epilogue to this children's book, Bruchac clears up the origins of the story. Longfellow relied on two sources: Henry Rowe Schoolcraft, an Indian agent to the Great Lakes area in 1822 and husband to Jane Johnston, granddaughter of a Chippewa chief; and Kah-ge-gah-ga-bowh (George Copway), a Chippewa lecturer and writer he had met in 1849.

92. Ibid.

93. Williams, "Bringing Honey to the Land of Milk and Honey," 34.

94. Ruth Marie Baldwin, *100 Nineteenth-Century Rhyming Alphabets in English*.

95. Ibid., 33.

96. Ibid., 66.

Chapter 4. After Bee Space

1. John Adams, *The Works of John Adams*, 345. The full quotation comes from a letter Adams wrote to William Tudor from Quincy, Mass., on August 11, 1818. In the letter to Tudor, Adams agrees with Judge George Minot's historical perspective that the Sugar Act was the precipitating event of the American Revolution: "General Washington always asserted and proved, that Virginians loved molasses as well as New Englandmen did. I know not why we should blush to confess that molasses was an essential ingredient of the American independence. Many great events have proceeded from much smaller causes." Molasses was categorized as a sugar, and the Sugar Act of 1764 signaled a change in British policy toward the colonies. The tax suggested that Britain would no longer pay for the colonization process, and the Sugar Act was quickly followed by the Stamp Act, the Quartering Act, and the Townshend Act; these taxes ignited the colonists' fury, and the eventual result was the American Revolution.

2. William Francis Allen, Charles Pickard Ware, and Lucy McKim Garrison, *Slave Songs of the United States*.

3. Ibid., 5.

4. Harold Courtlander, *Treasury of Afro-American Folklore*, 383.

5. Ibid., 442–43.

6. Ibid., 285.

7. Crane, *World History of Beekeeping and Honey Hunting,* 453.

8. See the following sources for information about Hetherington: Wyatt Mangum, "Quinby Smoker"; Frank C. Pellett, *History of American Beekeeping;* A. I. Root, "Editorial."

9. Mangum, "Quinby Smoker," 48.

10. "Obituary—O. O. Poppleton." After the war, Poppleton was in charge of establishing national cemeteries.

11. "War Stories: How Some Soldier Boys Managed a Bee Tree," *Courier-Journal,* November 23, 1887.

12. James Robertson, *Soldiers Blue and Gray,* 73.

13. Hattie Brunson (Mrs. C. G. Richardson), oral narrative, "Social Customs of the Past," Federal Writers' Project, June 28, 1938 (Beaufort, S.C.). Translated by Chlotilde R. Martin. Accessed on December 3, 2004, at http://memory.loc.gov/ammem/ndlpedu/features/timeline/civilwar.

14. Kent Masterson Brown, "Greenhorns and Honey Bees," 204.

15. *Civil War Trail in Arkadelphia.* Available at: http://www.arkadelphia.org.

16. Jason Wilson, personal e-mail communication, December 10, 2003.

17. Mason, "American Bee Books," 484.

18. Patterson, *Shaker Spiritual,* 427–28.

19. Goodwillie, *Shaker Songs,* 96.

20. See the following sources for information about Elder Henry C. Blinn: Brian Thompson, interview with the author, January 30, 2004; Deborah Burns, *Shaker Cities of Peace, Love, and Union;* Goodwillie, *Shaker Songs.*

21. Neal, *Kentucky Shakers,* 84. However, the Shakers were not the first to import bees to Kentucky. According to William Eaton's history of Kentucky beekeeping, Dr. N. P. Allen headed a committee of seven men who were interested in importing Italians in 1872: "The price of the queen bees was around 15.00 each" ("Sketches from Kentucky's Beekeeping History," 25).

22. John Wade, letter to Ruth Cox Wade, 1864. Available at: http://www.appalachianpower.com.

23. The title *Gleanings in Bee Culture* was shortened and has gone through several changes through the years. With the editor's permission, I use *Bee Culture*—its current title—for consistency.

24. Frank C. Pellett, *History of American Beekeeping,* 168.

25. A. I. Root, "Introduction," 1.

26. A. I. Root, "Our Latest Intelligence Corner," 35.

27. A. I. Root, "With Our Bees," 140.

28. Edward Goodell, "Bees by Sailing Ship and Covered Wagon," 39. Goodell's source is an anonymous man of Pennsylvania Dutch descent. Goodell did not provide a bibliography.

29. Oscar Haas, *History of New Braunfels and Comal County, Texas,* 218. The author is much indebted to the Sophienburg Museum, located in New Braunfels, which had a fine exhibit about the Naegelin Bakery and information about Naegelin's life.

30. Charles Abramson, "Charles Henry Turner," 643.

31. Joseph Dabney, *Smokehouse Ham,* 436.

32. Dadant, *The Life of C. P. Dadant.* M. G. Dadant writes that his grandfather happened to choose Hamilton, Illinois, because he had been corresponding with Mr. A. Morlot, who lived in Basco, Illinois. Morlot offered Charles Dadant a forty-acre farm at $20 an acre. The farm came with two or three colonies of bees.

33. Kent Louis Pellett, *Charles Dadant,* 14. Dadant did not support the Catholic Church in France or America. He refused to accept the Church's tolerance of poverty in France or slavery in the American South.

34. Ibid., 88.

35. Ibid., 115.

36. Ibid., 88. Dadant believed that Fourier's ideas could be successfully implemented into his business, but he wanted a slow evolutionary process. He had seen how the utopian experiments in France and the Brook Farm community in Massachusetts (where Nathaniel Hawthorne and Margaret Fuller were residents) had failed.

37. Ibid. It should also be remembered that Langstroth's great-grandparents were Huguenots and immigrated to America. His grandmother was a firm abolitionist, and this issue was another that both men agreed on.

38. Dadant had practice in overcoming his friends' religious convictions. His wife, Gabrielle Parisot, had reservations about Dadant's marriage proposal. She had been educated in a convent and never completely accepted her husband's socialism. She took great pride in her husband's accomplishments in the bee industry, however.

39. C.P. Dadant, *Life and Writings of Charles Dadant,* 53–57. A beekeeper named H.A. King asserted that Langstroth had not invented the moveable frame because there had been other hives available in France that were similar to Langstroth's. King had proceeded with his own patent using Langstroth's principles. Langstroth decided to sue King, and the letters championing both sides were aired in the *American Bee Journal.* The feud ended, however, when Dadant, who had experience with the French hives and knew the differences, wrote a letter gallantly titled, "Honor to whom Honor is due."

Dadant's letter was so logical and well-written that he was asked to become an editor of the journal shortly after the letter appeared in the magazine.

40. Pellett, *Charles Dadant,* 98–102.

41. Dumas, "Apiculture in Early Texas."
42. Ibid.
43. Frank C. Pellett, "Obituary—Dr. G. Bohrer," 123.
44. Crane, *World History of Beekeeping and Honey Hunting,* 590.
45. Prete, "Can Females Rule the Hive?" 136.
46. Dumas, "Apiculture in Early Texas."
47. Frank C. Pellett, *History of American Beekeeping,* 115.
48. A. I. Root, "Announcement—*Mrs. Tupper's Journal,*" 7.
49. Ibid.
50. Frank C. Pellett, *History of American Beekeeping,* 99.
51. Kent Louis Pellett, *Charles Dadant,* 70.
52. Ibid., 77. "Dadant observed ruefully, 'Woman varies.'"
53. "The Beekeepers Convention," *Pittsburgh Leader,* November 15, 1874.
54. A. I. Root, "Heads of Grain," 57.
55. Hubbell, *Book of Bees,* 24–25.
56. Ibid., 25.
57. A. I. Root, "She Will Be Here Today!" 149.
58. Dumas, "Apiculture in Early Texas," 107. Dumas does not say whether the wax flowers were made of beeswax.
59. Ibid. A fire around 1873 destroyed fences, mother cows, and grasses.
60. Ibid.
61. Donald B. Kraybill and Carl Bowman, *On the Backroad to Heaven.* The Hutterites began in 1528 and remain distinctive from the other Anabaptist German communities because they share everything.
62. Ibid.
63. Edward Stevenson, *Deseret News,* August 15, 1885. Edward Stevenson Papers, Latter-day Saints (LDS) Church Archives, box 8, folder 1.
64. Neal, *Kentucky Shakers,* 27.
65. A. I. Root, "Who's Who," 3.
66. Rufus Morgan, "Letters Written from San Diego County." See also Barbara Newton, "E.W. Morse." Newton's article details the trials and tribulations of Morse and his "Honey Adventure" after Rufus Morgan died. The 1879–80 drought was particularly stressful for flora and fauna in the region, and many beekeepers caught "Arizona fever" and left the region.
67. Ibid. See also Newton, "E. W. Morse, "153.
68. Joseph Moffett, *Some Beekeepers and Associates,* 31.
69. Eric Nelson, "History of Beekeeping in the United States," 3.
70. Ibid.
71. Ibid.

72. Frank C. Pellett, *History of American Beekeeping,* 36.

73. Harry H. Laidlaw Jr., *Contemporary Queen Rearing,* 159.

74. Ibid., 160.

75. Harry H. Laidlaw Jr. and Robert E. Page Jr., *Queen Rearing and Bee Breeding.*

76. See Laidlaw, *Contemporary Queen Rearing,* 170, and *Queen Rearing and Bee Breeding.* The terminology is a bit confusing, but a queenright hive is one that has a queen in it, as opposed to a queenless hive. A cell finisher is a colony used in the latter stages of queen rearing.

77. Frank C. Pellett, *History of American Beekeeping,* 109.

78. Ibid., 110.

79. Gordon Waller, "First Bees in Arizona," 435.

80. Ibid. Waller reports that "it has been documented that the Spanish introduced honey bees to Mexico and South America; but one should not conclude that the Spanish also brought them to Arizona since this is not consistent with the historical accounts of honey bees in Arizona."

81. Anti-Bee Monopolist, "A Stinger: The Trade in Bee Courses—A Protest against the Bee Convention—How the Bee Business Used to Be," *Pittsburgh Leader,* November 13, 1874, 1–2.

82. Ibid.

83. Ibid.

84. "All over the South," *Courier-Journal,* April 27, 1882, 5.

85. "Hiving Honey Bees: The Remarkable Conduct of Some Farmers' Boys and Girls Explained," *Courier-Journal,* July 30, 1882, 2.

86. Ibid.

87. "Bees in Church: A Swarm of Them Take Possession of a Maryland Chapel and the Congregation Suddenly Dissolves," *Courier-Journal,* September 24, 1888, 5.

88. Wallace Stegner, *The American West as Living Space,* 18.

89. Mark Twain, "How to Tell a Story," 239.

90. Dobie, "Honey in the Rock," 127.

91. Ibid., 129.

92. Ibid., 130.

93. Ibid., 132.

94. Ibid., 134.

95. Ibid., 135.

96. Herman Lehmann, *Nine Years Among the Indians.*

97. "Storming a Bee Castle," *Courier-Journal,* November 25, 1882, 3.

98. Welsch, "Funny Beesness," 36.

99. Mason, "American Bee Books," 350.

100. "Women as Bee Keepers," reprinted in *Courier-Journal,* April 12, 1891, 14.

101. Ibid.

102. Dumas, "Apiculture in Early Texas," 120.

103. Crane, *World History of Beekeeping and Honey Hunting*, 590.

104. Emily Dickinson, *The Poems of Emily Dickinson*, 392.

105. Ibid., 64.

106. Ibid., 430.

107. Ibid., 632.

108. Frank Stockton, *The Bee-Man of Orn.*

109. Baldwin, *100 Nineteenth Century Rhyming Alphabets in English,* 83.

110. "Bee Superstitions: Queer Customs in Honey Making Which Prevail in Parts of England; Similar Superstitions Also Exist in Several Parts of the Continent," *Courier-Journal,* August 17, 1890, 11.

111. Ibid.

112. Ibid.

113. Eaton, "Sketches from Kentucky's Beekeeping History," 27. In 1889, Reese and George W. Demaree contributed two exhibits to the World's Exposition in Paris, France—a solar wax extractor and Reese's original bee escape.

114. "Still Buzzing Away: The North American Bee-Keepers' Society at Lexington," *Courier-Journal,* October 8, 1881, 5.

115. "Utah Beekeepers Meeting" (*Journal History,* Salt Lake City, April 15, 1879), 2.

116. "Utah Beekeepers Meeting" (*Journal History,* Salt Lake City, April 10, 1890), 4.

117. Frank C. Pellett, *History of American Beekeeping,* 139–41.

118. Ibid., 20.

119. A. I. Root, *ABC and XYZ.*

120. E. F. Phillips, *A Brief Survey,* 45.

121. John Burroughs, *Birds and Bees,* 48.

122. Ibid., 85.

Chapter 5. Early Twentieth Century

1. Steven Stoll, *Fruits of Natural Advantage.* Stoll does not include a discussion of honey or honey bees in his fascinating work.

2. G. H. Cale, "Where Is Our Moses?" 7.

3. Kent Louis Pellett, *Charles Dadant,* 87. The issue of honey adultera-

tion produced a near estrangement between the two men. Pellett writes: "Dadant turned to the *ABJ* and lashed the Ohio editor. He claimed Root's knowledge of chemistry was faulty. He could not refrain, before laying aside his pen, from rapping the editor on his religious views and his frequent airing of them in [*Gleanings in*] *Bee Culture*." Pellett offered a warning in his biography: "Belligerent Dadant, ever ready to castigate those who fell afoul of his purposes."

4. Kim Flottum, "Bottom Board: Review of *Gleanings,*" 40.

5. Even though Hawaii wasn't a state, it too sent representatives to Washington to learn how to package their honeydew honey, an inferior honey but used in baking. Hawaiian beekeepers did not want to be accused of honey adulteration by California bakeries.

6. Frank C. Pellett, *History of American Beekeeping,* 119.

7. F. L. Aten, "The Past and Present of Progressive Beekeeping," *Texas Department of Agriculture Bulletin* 22 (Nov.–Dec. 1911): 353.

8. Moffett, *Some Beekeepers and Associates,* 67.

9. Rita Skousen Miller, *Sweet Journey,* 30.

10. "Beekeeping Grows in Beehive State," *Salt Lake Tribune,* January 13, 1967, F7.

11. Joe Graham, "100 Years Ago: Automobiles for Bee Work," 652.

12. Moffett, *Some Beekeepers and Associates,* 43.

13. Roger Morse, "Bee Tech," 12.

14. Ibid.

15. John Caldeira, "Texas Beekeeping History: The Southland Queen." Available at: http://Outdoorplace.org [accessed April 25, 2003].

16. Mrs. Booker T. Washington's maiden name was Margaret Murray.

17. Kim Flottum, "Bottom Board: Review of *Gleanings,*" 40.

18. Ibid.

19. Flottum, "Bottom Board: Anna Comstock," *Bee Culture* 125, no. 5 (1997): 62, 64.

20. H. D. Woods, "Beekeeping for Women," *Texas Department of Agriculture Bulletin* 22 (Nov.–Dec. 1911): 357–60.

21. Frank C. Pellett, "How the Women Win," 372–73.

22. Frank C. Pellett, "Honey Production in the Sage District," 296.

23. Beth Haiken, "Plastic Surgery and Beauty," 436.

24. Lois Banner, *American Beauty,* 216.

25. G. H. Cale, "Beeswax Needed in the War Effort," 247.

26. Aten, "The Past and Present of Progressive Beekeeping," 353.

27. Bodog F. Beck, *Honey and Health.* Bodok, a zealous campaigner for honey, reminded audiences that ice cream made with honey during this period "was a far superior product" (147).

28. "Honey in the Trenches of Europe," 839.
29. Cale, "Beeswax Needed in the War Effort," 247.
30. Haiken, "Plastic Surgery and Beauty," 443.
31. Ibid., 446. There were so many disabled British men that the government went to extensive lengths to entice American cosmetic surgeons and psychologists to help men regain morale, which officials had been worried would fall and thus threaten national security.
32. Jay Smith, "Thirty Five Years Ago," 188.
33. Philip A. Mason, "American Bee Books," 482. Quick's pamphlet was part of the Vocational Rehabilitation Series.
34. F. Eric Millen, "You Can If You Will," 158.
35. Everett Franklin Phillips, "Soldier Beekeepers," 302.
36. Millen, "You Can If You Will," 158.
37. C. P. Dadant, "Editor's Viewpoint" (1919), 371.
38. C. P. Dadant, "Editor's Viewpoint" (1920), 10.
39. Ibid., 11.
40. Gene Stratton-Porter, *The Keeper of the Bees,* 219, 193, 446.
41. H. F. Carrilton, "Letter to Editor," 89.
42. Moffett, "Mitchell Brothers of Missoula, Montana," 509; Moffett, "Powers Apiaries, Inc.," 201.
43. Miller, *Sweet Journey,* 114.
44. G. H. Cale, "Father Francis Jager," 115.
45. Frank C. Pellett, *History of American Beekeeping,* 178. Pellett argues that clover was found as early as 1739; recognized as valuable by Charles Dadant; and debated hotly by beekeepers throughout the late nineteenth century.
46. Ibid.,178–81.
47. Eaton, "Sketches from Kentucky's Beekeeping History," 30–31.
48. Charles Lesher, "North Dakota Sojourn—Part I," 156.
49. The first YMCA was established in 1850.
50. Devine, "The Bee as a Reformer," *American Bee Journal* 60, no. 3 (1920): 88. Although there may some prejudice against the Irish here, with the boy being called Mac and hailing from Boston, the writer clearly wants to show that bees were important in the juvenile reform system.
51. Julian Carter, "Birds, Bees, and Venereal Disease," 247.
52. Ibid.
53. Frank C. Pellett, "The Sweet Clover Belt of the South," 331.
54. Ibid., 332.
55. Ibid., 331.
56. Frank C. Pellett, "Mississippi," 405–7.
57. Greenville, Mississippi, was the most progressive city in Mississippi

during this time period. The powerful Percy family protected the interests of blacks and saw to it that they were educated and well treated, at least with respect to the standards of the rest of the state. Leroy Percy shrewdly saw that in order to have cotton, the city had to have labor—black labor. And to prevent black people from leaving, he improved their standard of living, as opposed to other plantation owners, who preferred violence.

58. Charles Abramson, "Charles Henry Turner," 643–44.

59. Ibid., 644. Turner was also very well known in the St. Louis area for his social causes.

60. Ibid.

61. A. I. Root, "Booker T. Washington."

62. Louis Wahl, "Some Experience with a Chinese Beekeeper," 432–33.

63. Bodog F. Beck, *Honey and Health,* 134.

64. May Berenbaum, *Buzzwords,* 92. "This bill not only remains on the books, it was strengthened in 1976 by a clause prohibiting the importation of eggs and semen as well as adult bees."

65. Randall Teeuwen, "Public Rural Education." It is worth noting that the Immigration Act passed in 1924 seriously restricted immigration from southern and eastern Europe, but state laws affecting immigrants had begun much earlier. According to Randall Teeuwen, World War I "revealed that a surprisingly high number of drafted men were unable to speak English. Consequently, many states, including Colorado (1918), passed legislation making it unlawful to teach German, and in some cases any foreign language, in private, parochial, and public schools" (23).

66. Eugene Shoemaker, "Memories of Bee Inspection in the Thirties and Forties, Part I," 69.

67. Frank Pellett, *History of American Beekeeping*, 97.

68. A. I. Root, *ABC and XYZ,* 405.

69. John M. Barry, *Rising Tide,* 399.

70. Jes Dalton, "Effects of the Great Flood," 199.

71. Henry Stabe, "Louisiana Flood Conditions," 233.

72. E. G. LeStourgeon, "Reconstruction in Louisiana," 8.

73. Ibid., 10.

74. Stabe, "Louisiana Flood Conditions," 336.

75. Robert Page Jr., "Obituaries: Harry Hyde Laidlaw Jr.," 58.

76. Jeffrey Steele, "American Indians in Nineteenth-Century Advertising," 46.

77. The company changed its name in 1964 to reflect a more correct pronunciation: Sue Bee Honey. The icon has also changed to a politically neutral honey bee.

78. Mooney, "Myths of the Cherokee," 309.

79. See C. Eric Lincoln, *The Black Muslims in America;* and Lynn Dumenil, *Modern Temper.*

80. Mary Kay Franklin, *Tribute to Walter T. and Ida Babin Kelley,* 13.

81. Walter Kelley, "Make It Easy," 12.

82. Eva-Marie Metcalf, "Invisible Child," 15–23.

83. Mason, "American Bee Books."

84. Ibid. Although Mason's list is extensive, two established women writers were Charlotte Maria Tucker, *Wings and Stings: A Tale for the Young* (1885); and Margaret Warner Morley, *The Bee People,* which appeared in 1899, and went on to be reprinted in 1900, 1901, 1911, 1923, and 1937. Before the stock market crash, *Historical Statistics of the United States* reports that the personal consumption of books was over 300 million, but it dropped immediately to under 300 million by 1929. Incidentally, sales of these items did not exceed 300 million again until 1943.

85. Black children's literature writers do not use the honey bee image, as far as I know. *The Brownies,* a children's literature magazine written especially for "children of the sun" beginning in 1920, does not contain any folklore or children's stories using bees, although a good number of Brer Rabbit stories appear, as do stories with other animals. Dianne Johnson-Feelings, *The Best of the Brownies Book;* W. E. B. DuBois, *Selections from "The Brownies Book."*

86. John Root, interview with the author, June 2, 2003.

87. Bodog F. Beck, "Christ with a Hive and Bees," 543. Crane, in *World History of Beekeeping and Honey Hunting,* also used the *Catholic Encyclopedia* as her reference, noting that the explanation has remained in use.

88. Crane, *World History of Beekeeping and Honey Hunting,* 600.

89. A. I. Root, *ABC and XYZ,* 428.

90. John Root, interview with the author, June 2, 2003.

91. Joe Graham, interview with the author, June 2003.

92. Tibor Szabo and Daniel Szabo, "Principles of Successfully Overwintering," 879.

93. Ibid., 877.

94. Bill Weaver and Mary Weaver, "How to Overwinter," 875.

95. Paul Salstrom, *Appalachia's Path to Dependency,* 96.

96. Charles Lesher, "North Dakota Sojourn—Part II."

97. Miller, *Sweet Journey,* 84.

98. Glenn Gibson, "Washington Scene: End of an Era," 475.

99. Roger Morse, "The Price of Honey," 366.

100. Cache County was the same place that N. E. Miller established his beekeeping operation.

101. Carol Edison, interview with the author, January 12, 2003.

102. Miller, *Sweet Journey,* 84.

103. Moffett, *Some Beekeepers and Associates,* 54.

104. Glen Perrins, "Honey Bees Pay College Fees!" 302.

105. John Bruce, "Adventures in Bee Inspection—Part I," 139.

106. Mason, "American Bee Books," 336. Mason provides the quote; he states that the quote comes from the preface of the book, page 9.

107. Robert Hemenway, *Zora Neale Hurston,* 231.

108. Zora Neale Hurston, *Their Eyes Were Watching God,* 183.

109. Richard Wright, "Between Laughter and Tears," 76.

110. Robert Page Jr., "Obituaries: Harry Hyde Laidlaw Jr.," 58.

111. Glen Stanley, "Iowa," 630.

112. Eugene Shoemaker, "Memories of Bee Inspection in the Thirties and Forties, Part IV," 234.

113. Bruce, "Adventures in Bee Inspection—Part I," 138.

114. Mary Louise Coleman, *Bees in the Garden and Honey in the Larder,* 28, 55, 89.

115. George Douglas, *Early Days of Radio Broadcasting*, 206.

116. Carolyn O'Bagy Davis, "Fannie and the Busy Bees," 102.

117. Ibid., 112.

118. Ibid.

119. Franklin, *Tribute to Walter T. and Ida Babin Kelley,* 16.

120. Cale, "Beeswax Needed in the War Effort," 247.

121. Ibid., 248.

122. "Bees and the War," *Newsweek,* September 27, 1943, 68.

123. Lawrence Goltz, "Retiring Editor of *Bee Culture,*" 40.

124. "Obituary—William Rowley James."

125. Moffett, *Some Beekeepers and Associates,* 41.

126. Ibid., 61.

127. Lawrence Goltz, "Obituary: Floyd Moeller," 440.

128. Wyatt Mangum, "Encounters with the Giant Honey Bee," 365.

129. Moffett, *Some Beekeepers and Associates,* 107.

130. Franklin, *Tribute to Walter T. and Ida Babin Kelley,* 16.

131. John Root, "News and Events: Leon A. Winegar."

132. Walter Kelley, Bee Supply Catalogue, 3.

133. Moffett, *Some Beekeepers and Associates,* 83.

134. Bonnie Sparks, interview with the author, Jan. 11, 2003.

135. John Bruce, "Adventurers in Bee Inspection—Part II," 179.

136. Clay Tontz, "California Dreaming," 101. A *gat* is an Americanized term for a machine gun, invented by Dr. Richard Gatling and first used in the Civil War.

137. Crane, *World History of Beekeeping and Honey Hunting,* 567.

138. Ibid., 567.

139. Karl von Frisch, *Dancing Bees,* 101.

140. Ibid., 117.

141. James Gould and Carol Grant Gould, *The Honey Bee,* 59.

142. Hubbell, *Book of Bees,* 93.

143. E. B. White, "Song of the Queen Bee," *New Yorker,* December 15, 1945, 37. I'm indebted to Sue Hubbell for bringing this poem to my attention in her wonderful book, *The Book of Bees . . . and How to Keep Them* (94).

144. Darlene Hine, William Hine, and Stanley Harrold, *African Americans: A Concise History,* 366.

145. Ibid., 303.

146. Robert Gordon, *Can't Be Satisfied*, 67.

147. Glenn Gibson, "Washington Scene: End of an Era," *Bee Culture* 144, no. 9 (1986): 475.

148. Bill Wilson, "45 Years of Foulbrood" (2000). Available at: http://www.beeculture.com.

149. Ibid.

150. Jane Black, "Enlisting Insects in the Military," *Business Week Online*, Nov. 5, 2001, http://web25.epnet.com (accessed Mar. 3, 2003).

151. Jerome Lawrence and Robert E. Lee, *Inherit the Wind,* 116.

Chapter 6. Late Twentieth Century

1. Bill Wilson, "45 Years of Foulbrood" (2000). Available at: http://www.beeculture.com.

2. David Halberstam, *The Fifties*, 142, 143.

3. Pamela Moore, "David and Goliath," 40.

4. Douglas Whynot, *Following the Bloom,* 82.

5. Walter Kelley, "Honey at Wholesale Prices," 107.

6. Walter Kelley, "Hard Times," 207.

7. Wilson, "45 Years of Foulbrood."

8. J. E. Eckert, *Rehabilitation of the Beekeeping Industry in Hawaii,* 22.

9. Stephen Petersen, "Alaska," 157.

10. Bob Koehnen, interview with the author, March 6, 2004.

11. "California Almonds: Technical Information" (Modesto, Ca.: Almond Board of California, 2004).

12. "Almonds: A Nutrition and Health Perspective" (Modesto, Ca.: Almond Board of California, 2003).

13. Roger Morse and Nicholas Calderone, *Value of Honey Bees as Pollinators,* 5–6.

14. Eric Mussen, interview with the author, June 15, 2004.

15. Ibid.

16. Yvonne Koehnen, interview with the author, March 6, 2004.

17. Bob Koehnen, interview, Mar. 6, 2004.

18. Lesher, "North Dakota Sojourn—Part II."

19. Clay Eppley, "The Rise and Fall of Beekeeping Along the Lower Rio Grande," 184.

20. Marla Spivak, interview with the author, June 23, 2004.

21. When a queen emerges from her cell, she will establish dominance by killing the emerging queens. African queens emerge before Italian queens and thus cannot be hybridized the way Kerr had hoped they could be.

22. Dewey Caron, *African Honey Bees in the Americas,* 11–12.

23. H. L. Maxwell, "Little Commercial Beekeeping in Virginia," 401. Emphasis Maxwell's.

24. Euclid Rains, *Count Me In,* 52.

25. Euclid Rains, *I'm Not Afraid of the Dark,* 93.

26. Clay Tontz, "The Mean Scarecrow," 773.

27. Walter Kelley, "Why Not Use Girls in the Honey House?" 305.

28. Joseph Moffett, "Mitchell Brothers of Missoula, Montana," 509.

29. Moffett, *Some Beekeepers and Associates,* 44.

30. Stephanie Coontz, *The Way We Never Were,* 110–13. Anthony Comstock was a moral reformer who managed to marshal support for a bill prohibiting the distribution of birth control information.

31. Joel M. Snow, "James 'Slim Harpo' Moore," *Blues Online*, July 16, 1996.

32. Gordon, *Can't Be Satisfied,* 197. In fact, the Rolling Stones got the name for their band from a Muddy Waters song.

33. Ibid., 163.

34. Ibid., 103.

35. Hine, Hine, and Harrold, *African Americans*, 379.

36. George Vest, "Beekeeping in Central Virginia," 400.

37. John Harbo, interview with the author, June 21, 2004.

38. Jesse Stuart, *The Beatinest Boy.*

39. John Root, "A *Gleanings* Interview: Karl Showler," 554–55.

40. Ibid.

41. Whynott, *Following the Bloom,* 80.

42. E. C. Martin, "Impact of Pesticides on Honeybees," 319.

43. Robert Nygren, "Honeybees and Microencapsulated Pesticides," 128–29.

44. Whynott, *Following the Bloom*, 27.

45. Franklin, *Tribute to Walter T. and Ida Babin Kelley*, 8.

46. Marianne Boruch, "Plath's Bees." Boruch gives an excellent analysis of Otto Plath's book, *Bumblebees and Their Ways*. "As a collection of straight facts on bees, or even as a how-to manual on beekeeping, Otto Plath's effort is a failure. It reads differently than most modern treatises on bees, which is to say, it thinks differently. . . . The book reads, in short, like field notes, carrying that sort of modesty and containment and absolute purpose" (4).

47. Sylvia Plath, *Ariel*, 65–66, 67–68.

48. Coontz, *The Way We Never Were*, 285. There is interesting discussion about the role of day care in America. When it first began at the turn of the century, it was associated with immigrant families, and thus could never receive the kind of funding that other countries enjoyed. Johnson's political party changed the status of day care by making it acceptable, thus, in parallel fashion, condoning the numbers of women working outside the home. Head Start is one of the few social programs that work. According to Stephanie Coontz, the Eisenhower Foundation has done several studies that found that school dropouts, teen pregnancies, and drug use dropped.

49. Sandra Beckett, ed., *Transcending Boundaries*, 50–51.

50. Lehman, interview with the author, June 29, 2004. See also Lehman, "Kids 'N Bees."

51. Carolyn Henderson, interview with the author, June 19, 2004.

52. Stephen Henderson, interview with the author, June 22, 2004.

53. Lawrence Goltz, "*The Swarm*," 381.

54. Lawrence Goltz, "*The Swarm* Gets Bad Reviews," 442.

55. Robert Heard, *Miracle of the Killer Bees*, 1–5.

56. Morse and Calderone, *Value of Honey Bees as Pollinators*, 124.

57. A. I. Root, *ABC and XYZ*, 124.

58. Whynott, *Following the Bloom*, 80.

59. Moffett, *Some Beekeepers and Associates*, 54.

60. Whynott, *Following the Bloom*, 81.

61. Franklin, *Tribute to Walter T. and Ida Babin Kelley*, 38.

62. "Obituary—Stanley F. Hummer."

63. Billy Hummer, interview with the author, June 23, 2004.

64. Chuck Anderson, interview with the author, March 26, 2004.

65. Charlotte Stevenson, "How *Not* to Move Bees," 281.

66. Clinton Heylin, *Can You Feel The Silence: Van Morrison A New Biography* (Chicago: Chicago Review, 2003), 23, 19.

67. Van Morrison, *Tupelo Honey* (audio recording, Exile, 1971).

68. See Matt Pommer, "Gag Order Doesn't Fit Dreyfus of Old," *Capitol Times*, April 22, 1993. I am grateful to librarian Ching Wong for supplying this article in a timely fashion.

69. A. I. Root, *ABC and XYZ,* 380.

70. T'Lee Sollenberger, "Family Bees," 128.

71. Susan Cobey, interview with the author, June 17, 2004. See also Malcolm Sanford, "Sue Cobey and Her New World Carniolans," *Bee Culture* 131, no. 1 (2003): 21–23.

72. Tom Webster, interview with the author, April 2, 2004.

73. Henry Wiencek, "Beehives," *Americana,* March/April 1981, 48.

74. Ibid., 47.

75. Carol Edison, interview with the author, January 15, 2003. Kathy Stephenson, "Desserts in the Desert," *Salt Lake Tribune,* July 24, 2002, B1, B3.

76. Richard Avedon, *In the American West.*

77. Elizabeth Royte, "Unfazed by All the Buzz," *Smithsonian* 33, no. 8 (2002), 25–27.

78. Hubbell, *Book of Bees,* 24, 26.

79. Ibid., 38, 47.

80. Fannie Flagg, *Fried Green Tomatoes at the Whistle Stop Cafe,* 110.

81. Ibid., 398.

82. Sidney Saylor Farr, *More than Moonshine.*

83. Jesse Stuart, *Men of the Mountains,* 227.

84. Ibid., 233.

85. Robert Morgan, "Moving the Bees," 43.

86. Stevie Ray Vaughan, "Honey Bee" (audio recording, Sony, 1984). *Couldn't Stand the Weather.* Rereleased in 1999.

87. Howard Scott, "Interview with Roxanne Quimby," 863.

88. Eric Mussen, interview with the author, June 15, 2004.

89. Alan Mairson, "America's Beekeepers: Hives for Hire."

90. Evan Sugden and Kristana Williams, "October 15: The Day the Bee Arrived," 18.

91. Ibid., 19.

92. Gene Kritsky, "Lessons from History," 370.

93. Marla Spivak, interview with author, June 23, 2004.

94. Erin Peabody, "SMR—This Honey of a Trait Protects Bees from Deadly Mites," 14.

95. Ibid., 15.

96. John Harbo, interview with the author, June 21, 2004.

97. Sue Hubbell, *Waiting for Aphrodite,* 136.

98. Cobey, interview with the author, June 17, 2004.

99. Tom Petty, "Honey Bee" (audio recording, Warner Bros., 1995).

100. Gloria Gaynor, "Honey Bee" (audio recording, MGM, 1974).

101. Tanya Long Bennett, "The Protean Ivy in Lee Smith's *Fair and Tender Ladies.*"

102. Lee Smith, *Fair and Tender Ladies,* 226.

103. Two items: first, Peter Fonda won the Florida State Beekeeper's award for his portrayal of Ulee. Second, his father, Henry Fonda, was a beekeeper in his own right, giving jars of Henry's Honey to family and friends.

104. Andy Jones, "Fonda's Golden Rule." Available at: http://www.roughcut.com (accessed January 12, 2001). Jones's article is a review of the movie.

105. *"Ulee's Gold"* (1997). Available at: http://www.beeculture.com. Jones, "Fonda's Golden Rule."

Epilogue

1. Whynott, *Following the Bloom,* 1. Whynott's argument is that contemporary beekeepers are very similar to nineteenth-century cowboys in terms of their travel, values, and important but overlooked services to the agricultural industry.

2. Mussen, interview with the author, August 5, 2004.

3. Andrew Revkin, "Bees Learning Smell of Bombs with Backing from Pentagon." Available at http://www.nyt.com.

4. Black, "Enlisting Insects in the Military," *Business Week Online.* Available at http://web25.epnet.com (accessed March 3, 2003).

5. "The Internest," *Economist,* no. 4 (2004): 78–79.

6. Gene Robinson, interview with the author, February 10, 2003.

7. Seth Borenstein, "Mistletoe," *Montreal Gazette,* A10.

8. Ibid.

9. Kim Kaplan, "What's Buzzing with the Africanized Honey Bee."

10. Marla Spivak, interview with the author, June 23, 2004.

11. Kim Flottum, interview with the author, May 1, 2003.

12. Jason Simpson, "Mobile Honey Production Unit Produced at Kelley's," *Leitchfield Record,* September 18, 2003.

13. Robin Mountain, "Rearing Your Own Kentucky Queen."

14. Stephen Henderson, interview with the author, June 22, 2004.

15. Marla Spivak, interview with the author, June 23, 2004.

16. Steven Stoll, *Larding the Lean Earth,* 217.

17. Stephenson, "Desserts in the Desert."

18. Allen Cosnow, "Use Correct Beekeeping Terms and Symbols." Cosnow, who argues that the continued use of the skep badly misleads the public, who "bless their adamantly ignorant hearts, know nothing at all about bees," then compares the skep to a pair of rusty pliers used by dentists: both tools should be obsolete in the public mind-set (89).

19. Sue Monk Kidd, *The Secret Life of Bees,* 141.

20. Ibid., 286.

Glossary

African honey bee or *Apis mellifera scutellata* (AHB): This honey bee began making its way to America in 1957 and is known for its aggressive temper. However, the bees are prolific honey producers.

American Bee Journal: Samuel Wagner started this periodical in 1861, but it was suspended because of the Civil War. This magazine was taken over by Charles Dadant and moved to Hamilton, Illinois. It addresses the topics of beekeeping, bee culture, and research.

American foulbrood (AFB): This bacterial disease affects the bees in the early larvae stage. The bacterial disease forms a spore that can remain alive for fifty years. The symptoms include bad odor, perforated brood cappings, and dead pupae.

Apis mellifera: Linnaeus originally assigned this name, which first qualified the domestic honey bee race as "honey-bearing."

Bee Culture: Begun by A. I. Root, this magazine offers advice and articles on bee management, culture, history, and so on.

Bee gum: Bees live in this log hive, generally taken from decayed black gum trees, hence the term.

Black bees or *Apis mellifera mellifera* (German black, British black, or French bee): This strain of honey bees was first brought to the colonies in the 1600s. These bees were hardy to cold temperatures, but were susceptible to foulbrood and ill-tempered to work with. They were the primary honey bees in the New World until Dadant, Wagner, Parsons, and Langstroth's efforts to import the Italian honey bee succeeded.

Brood: This term refers to bees in their stages of infancy—eggs, larvae, and pupae.

Carniolan bees, or *Apis mellifera carnica:* These bees were imported in 1877 by Charles Dadant. They never achieved the popularity of the Italian bees, although early importers thought these were the gentlest of the bees to work with.

Chalkbrood: This disease is caused by a fungus and is characterized by the chalky white appearance of dead larvae.

Drone: This male bee mates with the queen. The drones only last the summer season. Virgil called them "a pack of shirkers."

European foulbrood (EHB): This bacterial disease is characterized by dead unsealed brood, a sour odor, and watery consistency of dead brood.

Extractor: Major Francesco von Hruschka, who served in the Imperial Austrian army, invented this machine, which uses centrifugal force to pull the honey from the comb in a quick, easy fashion.

Honey: Bees make honey by gathering nectar from flowers and converting it into honey; used as food by bees; offers bees carbohydrates.

Honey Bee Act of 1922: This political act banned imports of different types of honey bees.

Italian or ("golden") bees, or *Apis mellifera ligustica:* This type of honey bee hails from the warmer climes of Italy. After the Civil War, many beekeepers began to import this type of bee because it was more resistant to American foulbrood and much gentler to work with than the German black bees.

Langstroth hive: This revolutionary hive was built according to bee space, which is the 3/8 inch of space between frames that bees need to build honeycomb. These frames can be removed when they are full of honey and wax without having to destroy the beehive.

Luting: This mixture of beeswax and tallow (bear or cattle fat) was used before the twentieth century (before paraffin became readily available) to waterproof or lubricate materials.

Nectar: This floral source provides a honey bee's carbohydrates.

Pollen: This floral substance provides the honey bee's protein needs.

Propolis: This substance is a mix of resins gathered from trees and beeswax. It is used by bees to fill in cracks in the hives and between extra spaces. As part of the bees' defense and insulation systems, propolis is extremely durable and difficult to crack.

Queen: This bee lays eggs and determines the sex of the eggs by the needs of the hive.

Round dance: This dance tells bees that the food sources are closer than ninety kilometers to the hive.

Royal jelly: This high-protein substance is fed to the young queen (which she will eat her entire life) and the larvae that are destined to become queens.

Skep: These rounded beehives were often made of rye straw; a few colonial towns hired a skeppist, to ensure a sufficient supply of honey and beeswax. Skeps were beneficial because they were moveable, easy, and cheap to fix, and they kept bees warm in the winter. But they were impractical for modern beekeeping because beekeepers had to destroy the hive in order to process the honey.

Small hive beetle: This pest arrived in the 1990s; bees are now developing resistance to its damage.

Smoker: Moses Quinby's smoker is the third major invention in beekeeping; smoke deadens the guard bees' sensory receptors and they will not attack; worker bees will eat more honey and therefore be more docile.

Squash bee, or *Peponapis pruinosa:* This solitary bee almost exclusively pollinates pumpkins and squash and is native to the New World. It is the subject of current research.

Suppressed mite reproduction (SMR): This is a genetic trait of honey bees. Researchers are trying to develop this trait among more honey bees.

Swarm: This natural response by the hive occurs when too many bees are inside. It is a healthy way of reproduction. The queen lays new queen eggs, and when the new queen emerges, the old queen will take a group of bees and leave for a new location. The young are left behind. Swarming is difficult for beekeepers, who will lose their investments if their bees fly away.

Tanging: This term refers to the taming of a bee swarm (usually in flight) by beating on pots and pans. It's an Old World custom that was perhaps started when a king demanded that people catching swarms should ring a bell, claiming the swarm as their own. Gradually, through the years, people forgot the king's decree and preferred to think that the noise is what convinced the swarms to alight on trees and be caught. The folk practice of tanging transferred to America.

Tracheal mite: This type of mite affects an adult honey bee's tracheal passages, making it difficult for it to breathe. The female mite feeds in the blood in the tracheal tube. It lays one male and several female eggs, and then dies. The male and females will mate inside the tracheal tube. These mites, then, cause damage to adult bees by impairing performance and piercing tracheal tubes.

Varroa mite: This pest first appeared in the United States from Asia in 1987 and did considerable damage to young bees. The varroa mite rides into the hive on an adult bee. Once in the hive, the varroa mite will bury itself in royal jelly. When the cell is capped, the mite will feed on the larvae. It lays one male egg and several female eggs. These will mate. When the young bee emerges, the varroa mite is ready to begin the cycle again. However, the bee has been damaged because it has been used as a host for the mite.

Waggle dance: This dance, performed by forager bees, describes to their sisters floral sources that are farther than ninety meters from the hive.

Wax moth: Although there are two kinds of wax moth—greater and lesser—

this book generally refers to greater wax moth, which is older. To paraphrase Roger Morse, wax moth destroys hives by laying larvae that destroy honeycombs by boring through the wax in search of food. Strong colonies can withstand wax moth by evicting the larvae, but especially in the South, bees are almost always dealing with this pest.

Worker bee: This type of bee spccializes in a number of tasks: foraging, cleaning the hive, taking care of the queen, guarding the hive, and converting pollen and nectar into honey.

DRAMATIS PERSONAE

Atchley, Mrs. Jennie: Nineteenth-century Texas beekeeper who edited *The Southland Queen.*

Bacon, Francis: This English philosopher wrote during the 1620s, comparing the unemployed masses to "swarms."

Bromenshank, Jerry: Twentieth-century researcher who does research with pollen detection of chemical weapons and electronic hive tracking devices.

Bruckish, Wilhelm: Nineteenth-century disciple of Johann Dzierzon, immigrates to New Braunfels, Texas, in the 1830s and begins beekeeping in the Texas cattle cradle.

Butler, Charles: Queen Elizabeth I's beekeeper and author of *The Feminine Monarchie* (1609) argued that there were three types of bees—the queen, her female workers, and the male drones. His book broke the traditional alliance with ancient Greek and Roman writers and focused on an English audience.

Cabet, Etienne: Nineteenth-century writer who wanted to create a utopia based on his novel, *A Voyage to Icaria.*

Cobey, Sue: Contemporary researcher who focuses on genetic breeding of New World Carniolans, a line of bees that shows mite resistance, winters well, and builds honey stores.

Cooper, James Fenimore: Eighteenth-century writer who popularized the bee hunter archetype.

Crèvecoeur, Jean Michael: Eighteenth-century French writer who compared the Nantucket society to a hive of bees. A bee hunter, he idealized the American society, often using bees as his social metaphor.

Dadant, Charles: Nineteenth-century French socialist immigrant to Illinois in 1865 who opened a bee supply company and edited the *American Bee Journal,* both of which are still active.

Dzierzon, Johann (last name pronounced *tziertzon*): Nineteenth-century German (Silesian) reverend and bee master; he was on the verge of discovering bee space and publishing his findings in German magazines. Samuel Wagner translates his work, only to meet Langstroth later and promote Langstroth's concept of bee space. Dzierzon and Langstroth became friends.

Eburne, Richard: Seventeenth-century English clergyman who encouraged his parishioners to go to the New World by using bee terms in his sermons.

Fonda, Peter: Son of a beekeeper (Henry), Peter starred in *Ulee's Gold* and accepted Florida Beekeeper of the Year on behalf of his character, Ulee Jackson.

Harbo, John: Contemporary researcher who studies suppressed mite resistance, a genetic trait that prevents varroa mites from reproducing.

Hartlib, Samuel: Seventeenth-century writer who wanted the English to use only honey as a sweetener and thus break England's reliance on sugar and slaves. He published a book about English beekeeping that was the first source about English conditions.

Hinkley, Gordon: President of the Church of Latter-day Saints during the 1970s; the use of the bee skep symbol tended to decline during his leadership.

Hruschka Major Francesco de: Member of Italian army who invented the honey extractor in 1865.

Hubbell, Sue: Twentieth-century beekeeper and writer.

Huber, Francois: Blind Swiss beekeeper who was in the process of discovering bee space. His writings help Langstroth think of the right distance and the moveable comb.

Irving, Washington: Nineteenth-century writer who wrote about honey bees on the Oklahoma frontier and chronicled the social uses of the word *bee* in New York society.

Kelley, Walter: Twentieth-century beekeeper who began his supply business in Louisiana, using bald cypress as his primary wood for hives. He moved his business to Kentucky in the 1930s and published *Modern Beekeeping*.

Keokuk: Nineteenth-century chief of the Sac and Fox Indians. He signed a treaty that allowed settlers into Wisconsin to the Mississippi River.

Laidlaw, Harry, Jr.: Twentieth-century researcher who pioneered artificial insemination and developed insemination instruments. His research furthered genetic testing with honey bees.

Langstroth, Lorenzo: Nineteenth-century descendant of French Huguenot grandparents. Langstroth invented the moveable frame hive, making commercialized beekeeping possible. He also imported Italian queens.

Mehring, Johannes: Nineteenth-century beekeeper who invented the honey extractor.

Owen, Robert: Nineteenth-century industrialist who wanted to create the perfect society, so he immigrated to American, founded New Harmony, and sold honey and beeswax as frontier commodities.

Pellett, Frank: Twentieth-century editor of the *American Bee Journal* and writer of *History of American Beekeeping* (1938).

Phillips, E. F.: Twentieth-century beekeeper who conducted workshops for World War I veterans.

Pratz, Antone le Page du: Eighteenth-century French overseer of a Louisiana sugar plantation.

Quinby, Moses: Nineteenth-century New York Quaker who invented the smoker in 1875.

Robinson, Gene: This researcher completed the Honey Bee Genome Sequence at Baylor University in 2003.

Root, A. I.: Nineteenth-century beekeeper who founded a bee supply company and the journal *Bee Culture.*

Root, Huber: Twentieth-century beekeeper who changed the Root Company's direction from bee supply to candle production.

Smith, Joseph: He selected the honey bee as the official icon of the Mormon (or Church of Jesus Christ of Latter-day Saints) religion.

Spivak, Marla: Contemporary researcher who studies hygienic behavior, a genetic trait that permits bees to detect diseases before they become infectious and to remove the affected area before the entire hive is infected.

von Frisch, Karl: Twentieth-century scientist who explained the round and waggle dances of bees. He won a Nobel Prize for his research.

Wagner, Samuel: Nineteenth-century beekeeper who translated Dzierzon's research into English, recognized the importance of Langstroth's discovery of bee space, and founded the *American Bee Journal* in 1861.

Young, Brigham: Joseph Smith's successor. When he successfully organized the Mormon Migration of 1847, he encouraged the use of bee skeps in daily life.

Zeisberger, David: Eighteenth-century Moravian missionary who traveled from Savannah, Georgia, to the Ohio frontier recording honey transactions and bee trees in his journals.

Bibliography

Abramson, Charles. "Charles Henry Turner: Contributions of a Forgotten African American to Honey Bee Research." *American Bee Journal* 143, no. 8 (2003): 643–44.

Adams, John. *The Works of John Adams, Second President of the United States.* Vol. 10. Boston: Little, Brown and Co., 1856.

Africans in America: America's Journey through Slavery. Dir. Orlando Bagwell. Alexandria, Va.: Public Broadcasting Service, 1998.

Allen, William Francis, Charles Pickard Ware, and Lucy McKim Garrison. *Slave Songs of the United States.* New York: Peter Smith, 1867.

"Almonds: A Nutrition and Health Perspective." Modesto, Calif.: Almond Board of California, 2003.

Amerman, Helen Adams. "Quilting Bee in New Amsterdam." *De Halve Maen* 53, no. 4 (1978): 5–6, 21.

Aten, F. L. "The Past and Present of Progressive Beekeeping." *Texas Department of Agriculture Bulletin* 22 (Nov.-Dec. 1911): 353.

Avedon, Richard. *In the American West.* New York: Harry N. Abrams, 1985.

Avery, Giles. B. "A Journal concerning Bees in the Second Order." Unpublished journal. Hancock Shaker Village, Pittsfield, Mass., 1851–54.

Bagster, S. *The Management of Bees.* London: Saunders and Outley, 1838.

Bailey, Rosalie Fellows. *Pre-Revolutionary Dutch Houses and Families in Northern New Jersey and Southern New York.* New York: Dover, 1968.

Baldwin, Ruth Marie. *100 Nineteenth Century Rhyming Alphabets in English.* Carbondale: Southern Illinois University Press, 1972.

Banner, Lois. *American Beauty.* New York: Knopf, 1983.

Barry, John M. *Rising Tide: The Great Mississippi Flood of 1927 and How It Changed America.* New York: Simon & Schuster, 1997.

Beck, Bodog F. "Christ with a Hive and Bees." *American Bee Journal* 80, no. 12 (1940): 543 and 546.

———. *Honey and Health.* New York: Robert M. McBride, 1939.

———. "The Great Seal of the State of Utah." *American Bee Journal* 81, no. 11 (1941): 497, 517.

———. "Tom Owen, the Mighty Honey Hunter." *American Bee Journal* 81, no. 4 (1941): 161–62.

Beckett, Sandra, ed. *Transcending Boundaries: Writing for a Dual Audience of Children and Adults.* New York: Garland, 1999.

Beisele, Rudolf Leopold. *The History of the German Settlements in Texas, 1831–1861.* Austin: Von Boeckmann-Jones, 1930.

Bennett, Tanya Long. "The Protean Ivy in Lee Smith's *Fair and Tender Ladies.*" *Southern Literary Journal* 30, no. 2 (1991): 76–95.

Berenbaum, May. *Buzzwords: A Scientist Muses on Sex, Bugs, and Rock 'n Roll.* Washington, D.C.: John Henry, 2000.

Bidwell, P. W. *A History of Northern Agriculture 1620 to 1840.* Washington: Carnegie Institution, 1925.

Black, Jane. "Enlisting Insects in the Military." *Business Week Online,* Nov. 5, 2001, http://web25.epnet.com (accessed Mar. 3, 2003).

Borenstein, Seth. "Mistletoe may have real effect on Birds and Bees." *Montreal Gazette,* Dec. 23, 2001. A10.

Boruch, Marianne. "Plath's Bees." *Parnassus: Poetry in Review* 17/18, no. 2/1 (1993): 76–96.

Bracht, Viktor. *Texas in 1848.* Translated by Charles Frank Schmidt. Manchaca, Tex.: German-Texas Heritage Society, 1849. Reprint, 1991.

Breininger, Lester. "Beekeeping and Bee Lore in Pennsylvania." *Pennsylvania Folklife* 16, no. 1 (1966): 34–39.

Bridenbaugh, Carl. *Vexed and Troubled Englishmen, 1590–1642.* New York: Oxford University Press, 1968.

Brophy, Alfred. "The Quaker Bibliographic World of Francis Daniel Pastorius's Bee Hive." *Pennsylvania Magazine of History and Biography* 122, no. 3 (1998): 241–91.

Brown, Kent Masterson. "Greenhorns and Honey Bees: The One Hundred and Thirty-Second Pennsylvania Volunteer Infantry at Antietam." *Lincoln Herald* 81, no. 3 (1979): 202–6.

Bruce, John. "Adventures in Bee Inspection—Part I." *American Bee Journal* 114, no. 3 (1986): 138.

———. "Adventurers in Bee Inspection—Part II." *American Bee Journal* 114, no. 4 (1986): 179–81.

Bruchac, Joseph. "Is *Hiawatha* an Authentic Native American Story?" In *Hiawatha and Megissogwon,* by Henry Wadsworth Longfellow. Washington, D.C.: National Geographic, 2001.

Brunson, Hattie (Mrs. C.G. Richardson). Oral narrative. "Social Customs of the Past." Federal Writers' Project, June 28, 1938 (Beaufort, S.C.). Translated by Chlotilde R. Martin. Accessed on December 3, 2004, at http://memory.loc.gov/ammem/ndlpedu/features/timeline/civilwar.

Burns, Deborah. *Shaker Cities of Peace, Love, and Union: A History of the Hancock Bishopric.* Hanover, N.H.: University Press of New England, 1992.

Burroughs, John. *Birds and Bees, Sharp Eyes and Other Papers.* New York: Houghton Mifflin, 1896.

Butler, Charles. *The Feminine Monarchie: Or a Treatise Concerning Bees and the Due Ordering of Them.* Oxford: Joseph Barnes, 1609. Reprints, 1623, 1634.

Cale, G. H. "Beeswax Needed in the War Effort." *American Bee Journal* 82, no. 6 (1942): 247–48.

———. "Father Francis Jager." *American Bee Journal* 81, no. 2 (1941): 115–16.

———. "Where Is Our Moses?" *American Bee Journal* 83, no. 1 (1943): 7.

"California Almonds: Technical Information." Modesto, Calif.: Almond Board of California, 2004.

Cannon, Hal. *The Grand Beehive.* Salt Lake City: University of Utah Press, 1980.

Carmony, Donald, and Josephine Elliot. "New Harmony, Indiana: Robert Owen's Seedbed for Utopia." *Indiana Magazine of History* 76 (1980): 161–261.

Caron, Dewey. *Africanized Honey Bees in the Americas.* Medina, Ohio: A.I. Root, 2001.

Carretta, Vincent. "Olaudah Equiano or Gustavus Vassa? New Light on an Eighteenth-century Question of Identity." *Slavery and Abolition* 20, no. 3 (1999): 96–105.

———, ed. *Unchained Voices: An Anthology of Black Authors in the English-Speaking World of the 18th Century.* Lexington: University Press of Kentucky, 1996.

Carrilton, H. F. "Letter to Editor." *American Bee Journal* 60, no. 3 (1920): 89.

Carter, Julian. "Birds, Bees, and Venereal Disease: Toward an Intellectual History of Sex Education." *Journal of the History of Sexuality* 10, no. 2 (2001): 213–49.

Cheney, Thomas, ed. *Mormon Songs from the Rocky Mountains: A Compilation of Mormon Folksong.* Vol. 53, *American Folklore Society.* Austin: University of Texas Press, 1963.

Cheshire, F. R. *Bees and Beekeeping.* Vol. 2. London: L. Upcott Gill, 1888.

Civil War Trail in Arkadelphia, The (web site). Jan. 26, 1998 (cited Feb. 20, 2003). Available from www.arkadelphia.org.

Cobbett, William. *A Year's Residence in the United States of America.* Fontwell, Britain: Centaur Press, 1964.

Cobey, Sue. Interview with author, June 17, 2004.

Coleman, Mary Louise. *Bees in the Garden and Honey in the Larder.* New York: Doubleday, Doran, 1939.

Cook, A. J. *The Bee-Keeper's Guide or Manual of the Apiary.* Lansing, Mich.: A. J. Cook, 1884.

Coontz, Stephanie. *The Way We Never Were: American Families and the Nostalgia Trap.* New York: Perseus, 1992. Reprint, 2000.

Cosnow, Allen. "Use Correct Beekeeping Terms and Symbols." *American Bee Journal* 143, no. 2 (2003): 89.

Courtlander, Harold. *A Treasury of Afro-American Folklore.* New York: Marlowe, 1996.

Crane, Eva. *The Archaeology of Beekeeping.* Ithaca: Cornell, 1983.

———. *The World History of Beekeeping and Honey Hunting.* New York: Routledge, 1999.

Crane, J. E. "Siftings." *Bee Culture* 44, no. 16 (1916): 714.

Crèvecoeur, John Hector. *Letters from an American Farmer.* Oxford: Oxford University Press, 1997.

Crosby, Alfred. *Ecological Imperialism: The Biological Expansion of Europe, 900–1900.* Cambridge: Cambridge University Press, 1986.

Dabney, Joseph. *Smokehouse Ham, Spoon Bread, & Scuppernong Wine: The Folklore and Art of Southern Appalachian Cooking.* Nashville, Tenn.: Cumberland House, 1998.

Dadant, C. P. "The Editor's Viewpoint: Are We Good Samaritans." *American Bee Journal,* November 1919, 371.

———. "The Editor's Viewpoint: Are We Good Samaritans." *American Bee Journal,* January 1920, 10–11.

———. *Life and Writings of Charles Dadant.* Dadant Library, Hamilton, Ill. Unpublished manuscript. N.d.

Dadant, M. G. *The Life of C. Dadant.* Hamilton, Ill., 1930.

Dalton, Jes. "Effects of the Great Flood." *Beekeepers Item* 11, no. 6 (1927): 199.

Davis, Carolyn O'Bagy. "Fannie and the Busy Bees." *Uncoverings: The Research Papers of the American Quilt Study* 23 (2002): 101–30.

De Jong, Gerald Francis. *Dutch in America, 1609–1974.* Boston: Twayne, 1975.

De Schweintz, Edmund. *The Life and Times of David Zeisberger: The Western Pioneer and Apostle of the Indians.* Philadelphia, Pa.: Lippincott, 1870. Reprint. Arno, 1971.

Devine, K. G. "The Bee as a Reformer." *American Bee Journal* 60, no. 3 (1920): 88–89.

Dickinson, Emily. *The Poems of Emily Dickinson.* Edited by R.W. Franklin. Cambridge, Mass.: Harvard University Press, 1998.
Dobie, J. Frank. "Honey in the Rock." In *Tales of Old-Time Texas,* 122–35. Edison, N.J.: Little, Brown, 1955.
Douglas, George. *The Early Days of Radio Broadcasting.* Jefferson, N.C.: McFarland Press, 1987.
DuBois, W. E. B. *Selections from "The Brownies Book."* Compiled by Herbert Aptheker. New York: Kraus-Thomason, 1980.
Dumas, Clark Griffith. "Apiculture in Early Texas." M.S. thesis, Southern Methodist University, 1952.
Dumenil, Lynn. *The Modern Temper: American Culture and Society in the 1920s.* New York: Farrar, Straus and Giroux, 1995.
Eaton, William. "Sketches from Kentucky's Beekeeping History." *Kentucky Academic Science* 26 (1965): 22–37.
Eckert, J. E. *Rehabilitation of the Beekeeping Industry in Hawaii.* Honolulu: University of Hawaii Agricultural Experiment Station, 1950.
Edward Stevenson Papers, Latter-day Saints (LDS) Church Archives, Salt Lake City, Utah.
Eppley, Clay. "The Rise and Fall of Beekeeping Along the Lower Rio Grande." *Modern Beekeeping* 37, no. 8 (1953): 183.
Ezell, Camp. *Historical Story of Bee County, Texas.* Beeville: Beeville Publishing, 1973.
Fabend, Firth Haring. *A Dutch Family in the Middle Colonies, 1660–1800.* New Brunswick: Rutgers University Press, 1991.
Farr, Sidney Saylor. *More than Moonshine: Appalachian Recipes and Recollections.* Pittsburgh: University of Pittsburgh Press, 1983.
Flagg, Fannie. *Fried Green Tomatoes at the Whistle Stop Cafe.* New York: Random House, 1987.
Flottum, Kim. "Bottom Board: Anna Comstock." *Bee Culture* 125 (1997): 64.
———. "Bottom Board: Review of *Gleanings.*" *Bee Culture* 125, no. 10 (1997): 40.
Franklin, Mary Kay. *Tribute to Walter T. and Ida Babin Kelley.* Clarkson, Ky.: Kelley, 1999.
Frisch, Karl von. *Dancing Bees: An Account of the Life and Senses of the Honey Bee.* New York: Harcourt Brace, 1952.
Garson, Robert. "Counting Money: The U.S. Dollar and American Nationhood, 1781–1820." *Journal of American Studies* 35, no. 1 (2001): 21–46.
Gaynor, Gloria. "Honey Bee." *Never Can Say Goodbye.* MGM, 1974.
Gibson, Glenn. "Washington Scene: End of an Era." *Bee Culture* 114, no. 9 (1986): 475.

Giraldi, Kathleen. "Noah Atwater and His Contribution to Westfield Life." *Historical Journal of Western Massachusetts* 3, no. 1 (1974): 42–50.

Goltz, Lawrence. "Obituary: Floyd Moeller." *Bee Culture* 106, no. 9 (1978): 440.

———. "Retiring Editor of *Gleanings in Bee Culture.*" *Bee Culture* 112, no. 1 (1984): 40–47.

———. "*The Swarm.*" *Bee Culture* 106, no. 8 (1978): 381.

———. "*The Swarm* Gets Bad Reviews." *Bee Culture* 106, no. 9 (1978): 442.

Goodell, Edward. "Bees by Sailing Ship and Covered Wagon." *Bee Culture* 97, no. 1 (1969): 38–40.

Goodwillie, Christian, ed. *Shaker Songs: A Celebration of Peace, Harmony, and Simplicity.* New York: Black Dog & Levanthal, 2002.

Goodwin, John. *The Pilgrim Republic.* Boston: Houghton Mifflin, 1920. Reprint. 1970. Kraus Press: New York.

Gordon, Robert. *Can't Be Satisfied: The Life and Times of Muddy Waters.* Boston: Little, Brown, 2002.

Gould, James, and Carol Grant Gould. *The Honey Bee.* New York: Scientific American Library, 1988.

Graham, Joe. "100 Years Ago: Automobiles for Bee Work." *American Bee Journal* 143, no. 8 (2003): 651–52.

Haas, Oscar. *History of New Braunfels and Comal County, Texas, 1844–1946.* San Antonio: Burke, 1968. Reprint, 1996.

Hahman, Frederick. "A History of Bee Culture in Pennsylvania." *American Bee Journal* 61, no. 10 (1933): 409.

Haiken, Beth. "Plastic Surgery and Beauty." *Bulletin of the History of Medicine* 69, no. 3 (1994): 436.

Halberstam, David. *The Fifties.* New York: Fawcett Publishing, 1993.

Hand, Wayland. "Anglo-American Folk Belief and Custom: The Old World's Legacy to the New." *Journal of the Folklore Institute* 7, no. 2/3 (1970): 136–55.

Harman, Ann. "Home Harmony." *Bee Culture* 116, no. 6 (1988): 350.

Hawke, David Freeman. *Everyday Life in Early America.* New York: HarperCollins, 1988.

Hayes, Bernie. "The Honey Bee and Napolean Bonaparte." *Bee Culture* 112, no. 11 (1984): 608.

Heard, Robert. *The Miracle of the Killer Bees: 12 Senators Who Changed Texas Politics.* Austin, Tex.: Honey Hill, 1981.

Hemenway, Robert. *Zora Neale Hurston: A Literary Biography.* Urbana: University of Illinois Press, 1980.

Heylin, Clinton. *Can You Feel The Silence: Van Morrison—A New Biography.* Chicago: Chicago Review, 2003.

Hine, Darlene, William Hine, and Stanley Harrold. *African Americans: A Concise History.* Upper Saddle River, N.J.: Prentice Hall, 2004.

Hogue, Charles. "Cultural Entomology." *Annual Review of Entomology* 32 (1987): 181–99.

"Honey in the Trenches of Europe." *Bee Culture* 44, no. 18 (1916): 839.

"How the Bees Saved America." *American Bee Journal* 57, no. 9 (1917): 307–8.

Hubbell, Sue. *The Book of Bees . . . And How to Keep Them.* Boston: Houghton Mifflin, 1988.

———. "The Honey War." In *On this Hilltop.* New York: Ballantine, 1997.

———. *Waiting for Aphrodite: Journeys into the Time before Bones.* Boston: Houghton Mifflin, 1999.

Hultgren, Roger, and Kathy Hultgren. "Old Swarming Devices." *Bee Culture* 116, no. 3 (1988): 158–60.

Hurston, Zora Neale. *Their Eyes Were Watching God.* New York: Harper & Row, 1937. Reprint, 1990.

Hurt, R. Douglas. *The Ohio Frontier: Crucible of the Old Northwest, 1720–1830.* Bloomington: Indiana University Press, 1996.

"The Internest." *Economist,* no. 4 (2004): 78–79.

Irving, Washington. *Knickerbocker's History of New York.* New York: Frederick Ungar, 1806. Reprint, 1928.

———. *A Tour on the Prairies.* Norman, Okla.: University of Oklahoma Press, 1962.

James, Hunter. *The Quiet People of the Land: A Story of the North Carolina Moravians in Revolutionary Times.* Chapel Hill: University of North Carolina Press, 1976.

Johnson-Feelings, Dianne, ed. *The Best of the Brownies Book.* New York: Oxford University Press, 1966.

Jones, Andy. "Fonda's Golden Rule: Review of *Ulee's Gold.*" July 18, 1997. Available at http://www.roughcut.com. (accessed January 12, 2001).

Jordan, Jerard, and Pat Jordan. *Spirit of America in Country Furniture 1700–1840.* Des Moines: Wallace-Homestead Book Co., 1975.

Jordan, Terry, and Matti Kaups. *The American Backwoods Frontier: An Ethnic and Ecological Interpretation.* Baltimore, Md.: Johns Hopkins University Press, 1989.

Kane, Thomas L. "The Mormons." Speech, Pennsylvania Historical Society, Pennsylvania, March 26, 1850.

Kaplan, Kim. "What's Buzzing with the Africanized Honey Bee." *American Bee Journal* 144, no. 5 (2004): 378–81.

Keefe, James, and Lynn Morrow. *The White River Chronicles of S. C. Turnbo: Man and Wildlife on the Ozarks Frontier.* Fayetteville: University of Arkansas Press, 1994.

Kelley, Walter. Bee Supply Catalogue. Clarkson, Ky.: Kelley Press, 1941.
———. "Hard Times." *Modern Beekeeping* 37, no. 9 (1953): 207.
———. "Honey at Wholesale Prices." *Modern Beekeeping* 37, no. 5 (1953): 107.
———. "Make It Easy." *Modern Beekeeping* 40, no. 1 (1956): 12.
———. "Why Not Use Girls in the Honey House." *Modern Beekeeping* 35, no. 9 (1950): 305.
Kidd, Sue Monk. *The Secret Life of Bees*. New York: Penguin, 2002.
Kishlansky, Mark. *A Monarchy Transformed: Britain 1603–1714*. London: Penguin, 1997.
Kraybill, Donald B. and Carl Bowman. *On the Backroad to Heaven: Old Order Hutterites, Mennonites, Amish, and Brethren*. Baltimore: Johns Hopkins University Press, 2001.
Kritsky, Gene. "Advances in Early Beekeeping." *American Bee Journal* 143, no. 4 (2003): 311–12.
———. "Beekeeping in Glass Jars." *American Bee Journal* 143, no. 8 (2003): 639–41.
———. "Castle Beekeeping." *American Bee Journal* 143, no. 1 (2003): 26.
———. "Langstroth and the Origin of the 'Bee Space.'" *American Bee Journal* 143, no. 10 (2003): 811–14.
———. "Lessons from History: The Spread of the Honey Bee in North America." *American Bee Journal* 131, no. 6 (1991): 367–70.
———. "Octagonal Hives." *American Bee Journal* 143, no. 11 (2003): 881–83.
Krochmal, Arnold and Connie. "Origins of United States Beekeeping." *American Bee Journal* 3 (1991): 162–63.
Kupperman, Karen Ordahl. "The Beehive as a Model for Colonial Design." In *America in European Consciousness: 1493–1750*, edited by Karen Ordahl Kupperman, 272–94. Chapel Hill: University of North Carolina Press, 1995.
———. "Death and Apathy in Early Jamestown." *Journal of American History* 66 (1979): 24–40.
———. *Facing Off: Encounters of English and Indian Cultures in America, 1640–1700*. Chapel Hill: University of North Carolina Press, 2000.
Laidlaw, Harry H. *Contemporary Queen Rearing*. Hamilton, Ill.: Dadant & Sons, 1979.
Laidlaw, Harry H., Jr., and Robert E. Page Jr. *Queen Rearing and Bee Breeding*. Cheshire, Conn.: Wicwas Press, 1997.
Lawrence, Jerome, and Robert E. Lee. *Inherit the Wind*. New York: Random House, 1955.
Le Sueur, Meridel. *North Star Country*. Minneapolis: University of Minnesota Press, 1945. Reprint, 1998.

Lehman, Kim. "Kids 'N Bees." *Bee Culture* 132, no. 3 (2004): 28–30.
Lehmann, Herman. *Nine Years Among the Indians, 1870–1879: The Story of the Captivity and Life of a Texan Among the Indians*. Translated by J. Marvin Hunter. 6th ed. Albuquerque: University of New Mexico Press, 1899. Reprint, 2000.
Lesher, Charles. "North Dakota Sojourn—Part I." *Bee Culture* 112, no. 3 (1984): 156–57.
———. "North Dakota Sojourn—Part II." *Bee Culture* 112, no. 4 (1984).
LeStourgeon, E. G. "Reconstruction in Louisiana." *Beekeepers Item* 12, no. 1 (1928): 1–10.
Lincoln, C. Eric. *The Black Muslims in America*. Boston: Beacon, 1961.
Linebaugh, Peter, and Marcus Rediker. *The Many-Headed Hydra: Sailors, Slaves, Commoners, and the Hidden History of the Revolutionary Atlantic*. Boston: Beacon, 2000.
Longfellow, Henry Wadsworth. "The Song of Hiawatha." In *Poems and Other Writings*, 141–279. New York: Penguin Putnam, 2000.
Longfellow, Henry Wadsworth and Jeffrey Thompson. *Hiawatha and Megissogwon*. Washington, D.C.: National Geographic, 2001.
Mairson, Alan. "America's Beekeepers: Hives for Hire." *National Geographic* 183, no. 5 (1993): 73–93.
Mancall, Peter C., ed. Envisioning America*: English Plans for Colonization*. Boston: Bedford St. Martins, 1995.
Mangum, Wyatt. "Encounters with the Giant Honey Bee: *Apis dorsata* Part I." *American Bee Journal* 144, no. 5 (2004): 365–67.
———. "The Quinby Smoker: It Changed Beekeeping Forever." *Bee Culture* 131, no. 6 (2003): 48–50.
Martin, E. C. "Impact of Pesticides on Honeybees." *Bee Culture* 106, no. 7 (1978): 318–20.
Mason, Philip A. "American Bee Books: An Annotated Bibliography of Books on Bees." Ph.D. diss., Cornell University, 1998.
Massey, J. Earl, ed. *Early Money Substitutes*. New York: American Numismatic Society, 1976.
Maxwell, H. L. "Little Commercial Beekeeping in Virginia." *American Bee Journal* 99, no. 10 (1959): 401.
McKinley, Daniel. "The White Man's Fly on the Frontier." *Missouri Historical Review* 58, no. 4 (1964): 442–51.
Merrick, Jeffrey. "Royal Bees: The Gender Politics of the Beehive in Early Modern Europe." *Studies in Eighteenth-Century Culture* 18 (1988): 7–35.
Metcalf, Eva Marie. "The Invisible Child in the Words of Tormod Haugen." *Barnboken* 1 (1992): 15–23.

Millen, F. Eric. "You Can If You Will." *American Bee Journal,* May 1919, 158.

Miller, Rita Skousen. *Sweet Journey: Biography of Nephi E. Miller.* Colton, Calif.: Miller Family Trust, 1994.

Moffett, Joseph. "Mitchell Brothers of Missoula, Montana." *Bee Culture* 109, no. 9 (1981): 508–9.

———. "Powers Apiaries, Inc., of Parma, Idaho." *Bee Culture,* April 1980, 201.

———. *Some Beekeepers and Associates.* Stillwater, Okla.: Frontier, 1979.

Monopolist, Anti-Bee. "A Stinger: The Trade in Bee Courses—a Protest against the Bee Convention—How the Bee Business Used to Be." *Pittsburgh Leader,* Nov. 13, 1874, 1–2.

Mooney, James. "Myths of the Cherokee." Washington, D.C.: Government Printing Office, 1898.

Moore, Pamela. "David and Goliath: Does Big Government Affect Small Beekeeping." *Bee Culture* 119, no. 1 (1991): 40–42.

Morgan, Robert. *Sigodlin: Poems.* Middletown, Conn.: Wesleyan University Press, 1990.

Morgan, Rufus. "Letters Written from San Diego County, 1879–1880." *Journal of San Diego History* 40, no. 4 (1994): 142–77.

Morrison, Van. *Tupelo Honey.* London: Exile, 1971.

Morse, Roger, ed. *The ABC & XYZ of Bee Culture.* 40th ed. Medina, Ohio: A. I. Root, 1990.

———. "Bee Tech." *American Heritage of Invention and Technology* 14, no. 4 (1999): 10–17.

———. "The Price of Honey." *Bee Culture* 106, no. 8 (1978): 366–67.

Morse, Roger, and Nicholas Calderone. *The Value of Honey Bees as Pollinators of U.S. Crops in 2000.* Medina, Ohio: A. I. Root, 2000.

Moseley, Rollin. "Why the Quilting 'Bee.'" *Bee Culture* 112, no. 12 (1984): 667.

Mountain, Robin. "Rearing Your Own Kentucky Queen." Frankfort: Kentucky State University, 2004.

Naile, Florence. *America's Master of Bee Culture: The Life of Lorenzo Langstroth.* Ithaca: Cornell University Press, 1976.

Neal, Julia. *The Kentucky Shakers.* Lexington: University Press of Kentucky, 1982.

Nelson, Eric. "History of Beekeeping in the United States." In *Agriculture Handbook*, no. 335 (1967): 2–4.

Newton, Barbara. "E.W. Morse and the San Diego County Beekeeping Industry, 1875–1884." *Journal of San Diego History* 27, no. 3 (1981): 150–60.

Nygren, Robert. "Honeybees and Microencapsulated Pesticides." *Bee Culture* 107, no. 3 (1979): 128–29.

"Obituary—O. O. Poppleton." *American Bee Journal,* 57, no. 12 (1917): 409.

"Obituary—Stanley F. Hummer." *American Bee Journal* 144, no. 1 (2004): 16.

"Obituary—William Rowley James." *American Bee Journal* 144, no. 1 (2004): 17.

Oertel, Everett. "Bicentennial Bees: Early Records of Honey Bees in the United States, Part I." *American Bee Journal* 116, no. 2 (1976): 70–71.

———. "Bicentennial Bees: Early Records of Honey Bees in the United States, Part II." *American Bee Journal* 116, no. 3 (1976): 114, 128.

———. "Honey Bees Taken to Cuba from Florida in 1764." *American Bee Journal* 121, no. 12 (1981): 880.

Olson, Ken. "Barreling Honey 100 Years Ago." *Bee Culture* 144, no. 10 (1986): 518.

Olson, Lester C. *Emblems of American Community in the Revolutionary Era: A Study in Rhetorical Iconology.* Washington, D.C.: London, 1991.

Online, The Handbook of Texas. *Old San Antonio Road.* University of Texas, [cited May 27 2004]. Available at: http://www.tsha.utexas.edu/handbook/online/articles/view/00/ex04.html.

Oxford English Dictionary (2nd) 2003 [cited Nov.18 2003]. Available from dictionary.oed.com.

Page, Robert, Jr.. "Obituaries: Harry Hyde Laidlaw Jr." *American Entomologist* 50, no. 1 (2004): 58–59.

Patterson, Daniel. *The Shaker Spiritual.* Princeton: Princeton University Press, 1979.

Peabody, Erin. "SMR—This Honey of a Trait Protects Bees from Deadly Mites." *Agricultural Research Service* (2004): 14–16.

Peckham, George. *A True Report of the Late Discoveries . . . by . . . Sir Humphrey Gilbert* (1583). *In Envisioning America: English Plans for the Colonization of North America, 1580–1640,* 62–70. Edited by Peter Mancall. New York: Bedford, 1995.

Pellett, Frank C. *History of American Beekeeping.* Ames, Iowa: Collegiate, 1938.

———. "Honey Production in the Sage District: Notes on the Methods of a Well-Known Beekeeper Who Produces Honey on a Large Scale." *American Bee Journal,* Sepember 1919, 296.

———. "How the Women Win: Three School Teachers Who Have Become Very Successful Beekeepers." *American Bee Journal,* November 1917, 372–73.

———. "Mississippi: Glimpses of Beekeeping and Some Other Things in the Land of Cotton, as Seen by the Associate Editor." *American Bee Journal,* December 1920, 405–7.

———. "Obituary—Dr. G. Bohrer." *American Bee Journal* 60, no. 3 (1920): 123.

———. "The Sweet Clover Belt of the South: Beekeeping in Alabama and Mississippi as Seen by Our Staff Correspondent on His Recent Trip." *American Bee Journal,* October 1917, 331–32.

Pellett, Kent Louis. *Charles Dadant: Bee Man from Champagne.* Hamilton, Illinois, 1927. Unpublished manuscript.

Perrins, Glen. "Honey Bees Pay College Fees!" *American Bee Journal,* August 1933, 302.

Petersen, Stephen. "Alaska: Beekeeping in the 49th State." *Bee Culture* 119, no. 3 (1991): 156–58.

Phillips, Everett Franklin. *A Brief Survey of Hawaiian Beekeeping.* Washington, D.C.: Government Printing Office, 1909.

———. "Soldier Beekeepers." *American Bee Journal,* September 1919, 302–3.

Pitzer, Donald, and Josephine Elliot. "New Harmony's First Utopians, 1814–1824." *Indiana Magazine of History* 75, no. 3 (1979): 225–300.

Plath, Sylvia. *Ariel.* New York: HarperCollins, 1961. Reprint, 1999.

Pommer, Matt. "Gag Order Doesn't Fit Dreyfus of Old," *Capitol Times,* April 22, 1993. Available at: http//:www.madison.com. (accessed Oct. 13, 2004).

Potter, Stephen. *Commoners, Tribute, and Chiefs: The Development of Algonquian Culture in the Potomac Valley.* Charlottesville, Va: University Press of Virginia, 1993.

Prete, Frederick. "Can Females Rule the Hive? The Controversy over Honey Bee Gender Roles in British Beekeeping Texts of the Sixteenth–Eighteenth Centuries." *Journal of the History of Biology* 24, no. 1 (1991): 113–44.

Rains, Euclid. *Count Me In.* Albertville, Alabama: Bimalto, 1999.

———. *I'm Not Afraid of the Dark.* Albertville, Alabama: Bimalto, 1996.

Raylor, Timothy. "Samuel Hartlib and the Commonwealth of Bees." In *Culture and Cultivation in Early Modern England: Writing and the Land,* edited by Michael Leslie and Timothy Raylor. Leicester, N.Y.: St. Martin's, 1992.

Revkin, Andrew. "Bees Learning Smell of Bombs with Backing from Pentagon," *New York Times*, vol. 151, no. 152771 (2002): A10.

Rink, Oliver. *Holland on the Hudson: An Economic and Social History of Dutch New York.* Ithaca: Cornell University Press, 1986.

Robertson, James I., Jr. *Soldiers Blue and Gray.* Columbia: University of South Carolina Press, 1998.

Root, A. I. *The ABC and XYZ of Bee Culture: An Encyclopedia Pertaining to the Scientific and Practical Culture of Honey Bees.* Edited by Roger Morse. Reprint; 40th ed. Medina, Ohio: A. I. Root, 1990.

———. "Announcement—*Mrs. Tupper's Journal.*" *Bee Culture* 2, no. 1 (1874): 7.

———. "Booker T. Washington." *Bee Culture* 44, no. 1 (1916): 39–40.

———. "Editorial." *Bee Culture* 28, no. 18 (1900): 738.
———. "Heads of Grain, from Different Fields—Anna Saunders Letter to the Editor." *Bee Culture* 2, no. 1 (1874): 57.
———. "Introduction." *Bee Culture* 1, no. 1 (1873): 1.
———. "Our Latest Intelligence Corner." *Bee Culture* 2, no. 3 (1874): 35.
———. "She Will Be Here Today!" *Bee Culture* 3, no. 12 (1875): 149.
———. "Who's Who." *Bee Culture* 3, no. 1 (1875): 3.
———. "With Our Bees." *Bee Culture* 3, no. 11 (1875): 140.
Root, John. "A *Gleanings* Interview: Karl Showler." *Bee Culture* 112, no. 10 (1984): 554–55.
———. "News and Events: Leon A. Winegar." *Bee Culture* 112, no. 9 (1984).
Rowe, Mike. *Chicago Blues: The City and the Music*. London, Eddison, 1973. Reprint. Da Capo, 1975.
Royte, Elizabeth. "Unfazed by All the Buzz." *Smithsonian* 33, no. 8 (2002).
Sachs, William. "The Business of Colonization." In *Studies on Money in Early America,* 3–14. New York: The American Numismatic Society, 1976.
Salisbury, Neal. "The Indians' Old World: Native Americans and the Coming of Europeans." In *American Encounters: Natives and Newcomers from European Contact to Indian Removal, 1500–1850,* 3–25, edited by Peter C. Mancall and James H. Merrell. New York: Routledge, 2000.
Salstrom, Paul. *Appalachia's Path to Dependency: Rethinking a Region's Economic History, 1730–1940*. Lexington: University Press of Kentucky. 1994.
Sanford, Malcolm. "Sue Cobey and Her New World Carniolans." *Bee Culture* 131, no. 1 (2003): 21–23.
Scott, Howard. "An Interview with Roxanne Quimby of Burt's Bees." *American Bee Journal* 143, no. 11 (2003): 863–68.
Scott, Kenneth. "A British Counterfeiting Press in New York Harbor, 1776." *The New York Historical Society Quarterly* 34, no. 2 (1955): 117–20.
"Sense and Nonsense: From *ABJ* 1866." *American Bee Journal* 100, no. 2 (1960): 71.
Seyffert, Carl. *Biene und Honig im Volksleben der Afrikaner* ("Beekeeping in Africa"). Leipzig, Germany: Voigtländer, 1930.
Sharpe, Kevin. *Politics and Ideas in Early Stuart England*. London: New York, 1989.
Shoemaker, Eugene. "Memories of Bee Inspection in the Thirties and Forties, Part I." *Bee Culture* 106, no. 2 (1978): 68–69.
———. "Memories of Bee Inspection in the Thirties and Forties, Part IV." *Bee Culture* 106, no. 5 (1978): 234–35, 241.
Simon, Carol. "Utah's Busy Beehives." *Smithsonian* 165, no. 4 (1981): 12.

Simpson, Jason. "Mobile Honey Production Unit Produced at Kelley's." *The Leitchfield Record,* Sept. 18, 2003.
Smith, Jay. "Thirty Five Years Ago." *Modern Beekeeping* 37, no. 8 (1953): 188.
Smith, Lee. *Fair and Tender Ladies.* New York: Ballantine, 1988.
Snow, Joel M. "James 'Slim Harpo' Moore." Blues Online. July 16, 1996. Available at: http//:physics.lunet.edu/blues/slim_harpo.html (accessed December 22, 2003).
Sollenberger, T'Lee. "Family Bees." *American Bee Journal* 143, no. 2 (2003): 127–28.
Stabe, Henry. "Louisiana Flood Conditions." *The Beekeepers Item* 11, no. 7 (1927): 233.
Stanley, Glen. "Iowa: Apiary Inspectors at the Heart of It." *Bee Culture* 115, no. 11 (1987): 630.
Steele, Jeffrey. "American Indians in Nineteenth-Century Advertising." In *Dressing in Feathers: The Construction of the Indian in American Popular Culture,* edited by Elizabeth Bird. Boulder, Colo.: Westview, 1996.
Stegner, Wallace. *The American West as Living Space.* Ann Arbor: University of Michigan Press, 1987.
———. *Mormon Country.* New York: Duell, Sloan & Pearce, 1942.
Stevenson, Charlotte. "How *Not* to Move Bees," *American Bee Journal* 143, no. 4 (2003): 281.
Stewart, L. R. "Early Midwest Beekeeping." *American Bee Journal* 81, no. 9 (1941): 412 and 421.
Stockton, Frank. *The Bee-Man of Orn and Other Fanciful Tales: New York: C. Scribner's Sons, 1887.*
Stoll, Steven. *Fruits of Natural Advantage: Making the Industrial Countryside in California.* Berkeley: University of California Press, 1998.
———. *Larding the Lean Earth: Soil and Society in Nineteenth-Century America.* New York: Farrar, Straus and Giroux, 2002.
Stratton-Porter, Gene. *The Keeper of the Bees.* Indianapolis: Indiana University Press, 1925. Reprint, 1991.
Stuart, Jesse. *The Beatinest Boy.* Edited by Jim Wayne Miller, Jerry Herndon and James Gifford. Ashland, Ky.: Jesse Stuart Foundation, 1980.
———. *Men of the Mountains.* Lexington: University Press of Kentucky, 1979.
Sugden, Evan, and Kristana Williams. "October 15: The Day the Bee Arrived." *Bee Culture* 119, no. 1 (1991): 18–21.
Szabo, Tibor, and Daniel Szabo. "The Principles of Sucessfully Overwintering Honey Bees." *American Bee Journal* 143, no. 11 (2003): 876–80.
Teale, E. W. "Knothole Cavern." *Natural History* 50, no. 10 (1942): 153–55.
Teeuwen, Randall. "Public Rural Education and the Americanization of the

Germans from Russia in Colorado, 1900–1930 (Part I)." *Journal of American Historical Society of Germans from Russia* 19, no. 1 (1996): 9–25.

Tontz, Clay. "California Dreaming." *Bee Culture* 117, no. 2 (1989): 100–102.

———. "The Mean Scarecrow." *American Bee Journal* 143, no. 10 (2003): 773.

Twain, Mark. "How to Tell a Story." In *Selected Shorter Writings of Mark Twain,* edited by Walter Blair. Boston: Houghton Mifflin, 1962.

———. *Roughing It.* Los Angeles: University of California Press, 1995.

Ulrich, Laurel Thatcher. *Good Wives: Image and Reality in the Lives of Women in Northern New England, 1650–1750.* New York: Vintage, 1991.

Unser, Daniel H., Jr. "The Frontier Exchange Economy of the Lower Mississippi Valley in the Eighteenth Century." In *American Encounters: Natives and Newcomers from European Contact to Indian Removal, 1500–1850,* edited by Peter C. Mancall and James H. Merrell. New York: Routledge, 2000. 216–39.

Van Wormer, Stephen. "Beeves and Bees: A History of the Settlement of Pamo Valley, San Diego County." *Southern California Quarterly* 68, no. 1 (1968): 37–64.

Vest, George. "Beekeeping in Central Virginia." *American Bee Journal* 99, no. 10 (1959): 400.

Wahl, Louis. "Some Experience with a Chinese Beekeeper: Cutting Bees out of a Side of a Building." *Bee Culture* 34, no. 4 (1906): 432–33.

Waller, Gordon. "First Bees in Arizona." *Bee Culture* 120, no. 8 (1992): 435.

Washington, Booker T., ed. *Tuskegee and Its People: Their Ideals and Achievements.* New York: Appleton and Co., 1905.

Watkins, Lee. "John S. Harbison: California's First Modern Beekeeper." *Agricultural History* 43, no. 2 (1969): 239–48.

Weaver, Mary, and Bill Weaver. "How to Overwinter Small Hives Successfully." *American Bee Journal* 143, no. 11 (2003): 872–75.

Webb, Walter Prescott. *The Great Plains.* New York: Grosset & Dunlap, 1931.

Weinlick, John R. *Count Zinzendorf: The Story of His Life and Leadership in the Renewed Moravian Church.* Nashville: Abingdon, 1956.

Welsch, Roger. "Funny Beesness." *Bee Culture* 116, no. 1 (1988): 36.

White, E. B. "Song of the Queen Bee." *New Yorker,* December 15, 1945, 37.

Whynott, Douglas. *Following the Bloom: Across America with the Migratory Beekeepers.* New York: Penguin, 1991.

Wiencek, Henry. "Beehives." *Americana* (March/April 1981): 46–48.

Williams, Catherine. "Bringing Honey to the Land of Milk and Honey: Beekeeping in the Oregon Territory." *American West* 12, no. 1 (1975): 32–37.

Wilson, Bill. *45 Years of Foulbrood* (128.10) 2000 (cited October 15, 2003). Available from www.beeculture.com.

Withington, Ann Fairfax. "Republican Bees: The Political Economy of the Beehive in Eighteenth-Century America." *Studies in Eighteenth-Century Culture* 18 (1988): 39–77.

Wood, William. *New England's Prospect.* Vol. 68. *The England Experience.* Amsterdam: Da Capo, 1634. Reprint, 1968.

Woodham-Smith, Cecil. *The Great Hunger: Ireland 1845–1849.* London: Penguin, 1962. Reprint, 1991.

Woods, H. D. "Beekeeping for Women." *Texas Department of Agriculture Bulletin* 22 (November–December 1911): 357–60.

Wright, Richard. "Between Laughter and Tears. Review of Their Eyes Were Watching God." In *Critical Essays on Zora Neale Hurston*, 77–78, edited by Gloria L. Cronin. New York: G.K. Hall & Co., 1998.

Wuorinen, John. *The Finns on the Delaware, 1638–1655.* New York: Columbia University Press, 1938.

Youngs, Isaac Newton. *Journal: Tour with Brother Rufus Bishop Through the State of Ohio and Kentucky.* Old Chatham, N.Y.: Shaker Museum and Library, 1834.

Zeisberger, David. *Diary of David Zeisberger: A Moravian Missionary Among the Indians of Ohio.* Translated by Eugene E. Bliss. Vol. 1. Cincinnati: Robert Clark, 1885. Reprint, 1972.

PERMISSIONS

Permission is gratefully acknowledged for the following:

Fourth and seventh stanzas from "The Arrival of the Bee Box" from *Ariel* by Sylvia Plath. Copyright © 1963 by Ted Hughes. Reprinted by permission of HarperCollins Publishers Inc. and Faber and Faber Ltd.

Excerpt from "The Beatinest Boy" by Jesse Stuart, copyright © 1980. Used by permission of Jesse Stuart Foundation.

Excerpt from *A Book of Bees* by Sue Hubbell. Copyright © 1988 by Sue Hubbell. Reprinted by permission of Houghton Mifflin Company. All rights reserved.

Excerpt from *Fried Green Tomatoes at the Whistle Stop Café* by Fannie Flagg, copyright © 1987 by Fannie Flagg. Used by permission of Random House, Inc.

Excerpt from "Honeybee" written by Melvin Steals, Mervin Steals, and Matt Ledbetter. Performed by Gloria Gaynor. Copyright © 1973 (renewed 2001). Used by permission of EMI Sosaha Music, Inc. and Jonathan Three Music.

Excerpt from "Honey Bee" written by Muddy Waters, copyright © 1959, 1987 Watertoons (BMI)/Administered by BUG. All rights reserved. Used by permission.

Excerpt from "Honey Bee" written by Stevie Ray Vaughan, copyright © 1985 Ray Vaughan Music (ASCAP)/Administered by BUG. All rights reserved. Used by permission.

Excerpt from "Honey Bee" by Tom Petty, copyright © 1994 Gone Gator Music (ASCAP). Administered by WB Music Corp. (ASCAP). All rights reserved. Used by Permission of Warner Brothers Publications, U.S., Inc., Miami, FL 33014.

"I'm a King Bee." Words and music by James H. 'Slim Harpo' Moore. Copyright © 1957 (renewed) by Embassy Music Corporation (BMI). International copyright secured. All rights reserved. Reprinted by permission.

INDEX

References to figures are in boldface type.